Rays of Hope

The Transition to a Post-Petroleum World

RAYS of HOPE

The Transition to a Post-Petroleum World

DENIS HAYES

A Worldwatch Institute Book

W · W · NORTON & COMPANY · INC ·
NEW YORK

Library of Congress Cataloging in Publication Data

Hayes, Denis, 1944–
  Rays of hope.

  Includes bibliographical references and index.
  1. Power resources.   2. Solar energy.   3. Atomic
energy.   I. Title.
HD9502.A2H37   1977   333.7   77–23037
ISBN 0–393–06418–2
ISBN 0–393–06422–0 pbk.
    4 5 6 7 8 9 0

To my mother, Antoinette S. Hayes

Contents

Foreword

THE PROJECTED PEAKING and subsequent decline in world production of petroleum, now humanity's principal source of commercial energy, is only half a generation away. Some of the more durable cars being bought today will still be in use when the oil production downturn begins. The transition to a world with dwindling oil output is an imminent reality. It could be a painful transition if we do not prepare for it.

The question is not whether we make the transition or not. We will make it. The only question is whether it will be a smooth one, the result of careful planning and preparation, or chaotic, the result of a succession of worsening economic and political crises. Few, if any, national leaders have any vision of what their societies will look like in a post-petroleum world. Although we might prefer to leave the adjustment to subsequent generations, history will not have it so. It has bequeathed to our generation the responsibility for planning and making the transition.

The oil production curve for the United States can serve as a prototype for the world's, underlining the inevitability of a global downturn. After decades of growth, U.S. oil production peaked in 1970. It has declined each year since. A similar downturn in world oil production is projected for 1990 or shortly thereafter, but there is one important difference. While the United States could turn to other countries to fill its oil deficit, the world as a whole cannot.

Knowledge that the world would eventually run out of petroleum has not been an urgent concern until recently because nuclear power was expected to fill the void. But the nuclear dream is beginning to fade as atomic power generates new economic, ecological, and political problems. *Rays of Hope* attempts to think through some of the steps which must be taken in energy conservation and in developing alternate sources of energy. It looks at the energy problem in a global perspective, recognizing that the firewood crisis in the Third World and overconsumption of energy in gas-guzzling private automobiles in the affluent countries intersect in the world petroleum market. Humanity now faces one of the most momentous adjustments in modern history, with little time to prepare for it. In the first instance, the transition is technological, but it promises to reshape our economic system and social structures as well. Denis Hayes' analysis suggests that a world which comes to depend heavily on renewable energy sources will be far different from the one in which we now live. As solar energy, both direct and indirect, expands in importance, it is certain to affect the distribution of population between countryside and city and possibly even the ultimate population carrying capacity of the planet.

Rays of Hope is an early effort to explore the shape of the post-petroleum world and how we get from here to there. The book's great strength is its perspective, historical and global. Denis Hayes helps opinion leaders and decision-makers at all levels to see how the energy problem will become the energy crisis if action is not taken quickly.

Hayes was the coordinator of the first Earth Day in 1970. He has been a Visiting Scholar at the Smithsonian Institution, and more recently he served as director of the state Energy Office in Illinois. His experience with environmental and energy issues and his skills as an analyst have been bolstered by travel in France, Saudi Arabia, India, Australia, and Japan, where he discussed national energy strategies and alternative energy sources with political leaders and energy experts.

This book is part of a much broader effort by the Worldwatch Institute to identify and focus public attention on emerging global problems. Certainly the transition to a post-petroleum world must rank high on any such list. *Rays of Hope*, the second Worldwatch book, follows *Losing Ground: Environmental Stress and World Food Prospects*, by Erik Eckholm (W. W. Norton, 1976). Portions of it were published in Worldwatch Paper 4, "Energy: The Case for Conservation"; Worldwatch Paper 6, "Nuclear Power: The Fifth Horseman"; Worldwatch Paper 11, "Energy: The Solar Prospect"; *BioScience; Natural History;* and the New York *Times.*

LESTER R. BROWN,   *President*
Worldwatch Institute

Acknowledgments

LIKE MOST MODERN nonfiction books, this one is not the product of a single intellect. My debts are numerous, large, and widespread.

I was particularly fortunate to have the full support of the Worldwatch Institute in this effort. Lester Brown's unfailing enthusiasm, broad experience, and stimulating ideas were of inestimable value throughout the project. The entire manuscript was critically reviewed by Erik Eckholm, Patricia McGrath, Kathleen Newland, Frank Record, Linda Starke, and Bruce Stokes, and their many helpful comments were gratefully incorporated. My editor, Kathleen Courrier, clarified my prose and helped establish a coherent structure for the manuscript. Frances Hall's fast and accurate typing kept the work proceeding smoothly, while Blondeen Duhaney, Marion Frayman, and Trudy Todd helped with some of the secretarial load.

Many outside reviewers helped shape portions of the manuscript. Various chapters relating to energy supplies profited from reviews by Wilson Clark, Dean Abrahamson, M. King Hubbert, Chester Cooper, Carlos Stern, David Comey, and Jim Benson. Frank von Hippel provided an invaluable critique of the chapter on nuclear power, while the comments of Alan Poole, Roscoe Ward, and the reviewers for *Bio-Science* magazine strengthened the chapter on organic energy sources.

The chapters on energy use were improved by the suggestions of Erik Hirst, Joel Darmstadter, Lee Schipper, and Clark Bullard. Many of the economic portions of the manuscript were scrutinized by C. Fred Bergston, Herman Daly, and/or Talbot Page, while Leon Lindberg and

David Orr assessed the final draft through the lenses of political scientists.

During a trip around the world to gather material for this book, I received invaluable assistance from a large number of people. Among the most helpful were Paulo Krahe, Charles Watson-Munro, John Price, Neal Barrett, J. J. Kowalczewski, Prince Saud al-Faisal, Prince Mohammed al-Faisal, Prince Turki al-Faisal, Sheik Ahmed Zaki Yamani, Farouk Akdar, Masao Kunihiro, Jean Robert, R. C. Bhargava, Arjun Makhijani, H. R. Srinivasan, M. C. Gupta, R. B. Ajgaonkar, C. R. Das, M. K. Gopalakrishnan, Tom Mathew, Mans Lonnroth, and Lars Josephson.

The general orientation and thrust of the book owes much to searching conversations over the years with Bruce Hannon, Amory Lovins, John Holdren, Grant Thompson, Jeremy Stone, and Sam Love. Every section of the manuscript profited from the detailed and thoughtful comments of my wife, Gail Boyer Hayes. Any remaining errors of fact or judgment are my responsibility alone.

D. H.

Worldwatch Institute
1776 Massachusetts Ave., N.W.
Washington, D.C. 20036
U.S.A.

Rays of Hope

The Transition to a Post-Petroleum World

1. Introduction: Twilight of an Era

For RICH LANDS and poor alike, the energy patterns of the past are not prologue to the future. The oil-based societies of the industrial world cannot be sustained and cannot be replicated; their spindly foundations, anchored in the shifting sands of the Middle East, have begun a long, irreversible process of erosion. The agrarian world's reliance upon firewood has proved similarly precarious as forests recede and even disappear entirely. Although the oil crisis dominates the headlines, hundreds of millions are affected by the shortage of firewood.

Until recently, most poor countries eagerly looked forward to entry into the oil era, with its ubiquitous automobiles, airplanes, and diesel tractors. However, the recent fivefold increase in oil prices virtually guarantees that the Third World will never derive most of its energy from petroleum. For two decades, the rich countries have proceeded on the belief that the oil era would be superseded by the nuclear age. However, it now appears increasingly unlikely that nuclear power will ever become the industrial world's principal source of commercial energy.

The entire world thus stands at the edge of an awesome discontinuity in its production and use of energy. The range of possible energy options is narrowed by factors other than just the scarcity of certain fuels. Long before all the earth's coal has been burned, for example, coal use may be halted by the impact of the rising atmospheric carbon dioxide levels on climate. Solar energy will not run out for 10 billion years, but some solar technologies will be limited by a scarcity of the materials needed to build devices to capture and store the energy in sunlight.

In both the Third World and the industrial world, various physical limits on energy growth have begun to assert themselves. Mountains are denuded by scavengers in a desperate quest for firewood, and ever-hungry draft animals have little surplus energy for tilling the fields. The growing demands of an expanding population push traditional energy systems past their carrying capacities—leading in some cases to ecological collapse. In the developed nations, a lack of water in the American West, a scarcity of suitable land in the Netherlands, and a lack of healthful air over much of Japan have all acted as brakes on energy growth.

In addition to such physical limits, energy supplies are also influenced by social factors. Despite the best efforts of powerful supporters in all quarters, energy growth is already pressing against social limits in much of the industrial world. Farmers are opposing strip mines; environmentalists are fighting petroleum refineries; and skyrocketing construction costs have led to the cancellation of plans for many nuclear reactors.

Every energy source is under the heels of both physical and social constraints. Some such limits are absolute—when natural gas runs out, natural gas consumption must stop—but more often they manifest themselves as increasingly severe hindrances on growth. Depending upon the mix of technologies employed, different types of constraints will come into play, but at some point accumulated constraints will halt further energy growth completely.

Heat: The Ultimate Limit

The earth has passed through many climatic epochs, ranging from ice ages to ice-free ages. The global climatic system appears to be delicately balanced; rather small alterations can trigger vast changes because certain basic physical processes can accelerate the effects of a perturbation. For example, ice and snow tend to reflect sunlight instead of absorbing it as heat. When an outside heat source melts the ice and snow on the ground, both the runoff and the bare ground itself absorb additional heat from the sun, melting still more ice and snow. Because small events appear capable of causing large climatic changes—some of which may be irreversible on any time scale of interest—even small changes must be executed with utmost caution.[1]

The constant flow of power from the sun, averaged over the surface of the rotating earth, amounts to about 340 watts per square meter. More than half this sunlight is reflected and scattered by clouds and airborne particles, so the earth's surface finally absorbs about 160 watts per square meter. Energy use by human beings now totals less than one ten-thousandth of the solar influx, and the global heat impact of this level of use seems to be negligible. The local effects of human energy use are sometimes quite significant, however.

Electrical power plants, industrialized cities, and various other energy-intensive sites each radiate several times more heat than they receive from the sun. Such "hot spots" affect local weather; they can help determine the frequency of snow, hail, thunderstorms, and even small tornadoes. Consequently, the number of energy facilities that can be built in any one area must be limited. However, the direct thermal effects of human energy use do not appear to be a cause for global concern unless such use increases severalfold above its current level.

Carbon dioxide (CO_2), a by-product of all fossil fuel combustion, poses a greater problem. Adding CO_2 to the air raises the earth's temperature by retarding the radiation of heat into space—a phenomenon known as the "greenhouse" effect. Since CO_2 can linger in the atmosphere for hundreds or perhaps thousands of years, the impact of CO_2 emissions is cumulative. Total atmospheric carbon dioxide has increased at least 10 percent in the last three-quarters of a century. Quite probably, future fossil fuel consumption will be limited by atmospheric tolerance for carbon dioxide long before the world fossil resource base has been exhausted. Between 1900 and 1975, CO_2 emissions grew from 2,000 million to 18,000 million tons per year. In late 1976, the Scientific Committee on Problems of the Environment, a leading independent group of international environmental experts, reported that it considered atmospheric CO_2 to be the world's foremost environmental problem.

Particulates, bits of matter so small that they can remain suspended in the air for lengthy periods, present another environmental problem. Though many natural processes produce particulates, fuel combustion is thought to account for about one-third of the total created annually. Particulates are believed to counteract the warming effects of carbon dioxide by reflecting incoming sunlight back out to space, and by increasing the density of cloud cover. But calculations about the net effect of such phenomena are rife with uncertainty.

In the popular media, it is often asserted that the cooling effect of particulates and the warming effect of CO_2 are balancing one another out. The implication is that we therefore have no cause for worry. But even if some such balance exists, it will almost certainly be upset eventually by the fundamental differences in the distribution and longevity of the two substances.

Any balance between the effects of carbon dioxide and those of particulates is delicate indeed. Carbon dioxide is circulated around the world's atmospheric system, while particulates blanket only the Northern Hemisphere. The global north is experiencing a cooling trend, while the Southern Hemisphere is simultaneously warming up—bearing out the "greenhouse" hypothesis.[2] Moreover, CO_2 will remain in the atmosphere much longer than particulates; to the extent that particulates temporarily hide the long-term warming effects of CO_2, they may prompt us to allow fuel use to exceed a level that informed prudence might dictate.

Climatic problems are incredibly complex. Before all the variables are entirely understood, human energy use could trigger far-reaching consequences. A decision to retard the rate of energy growth would reduce the chance of making a dreadful mistake. Such a decision would have to be made in the face of much uncertainty, but the consequences of not doing so could prove irreversible.

Pollution—Troubled Waters

All conventional energy sources—even the so-called "clean" ones like natural gas and geothermal power—generate pollution. As the use of such sources increases, the problems of pollution control grow more formidable. While a 90 percent effective control might be sufficient for a small source of pollution, a 99 percent effective control may become necessary when that source grows tenfold. But the incremental costs of each additional degree of control increase disproportionately; to capture the last few percent often costs many times as much as to capture the first 90 percent.

The world's experience with oceanic oil illustrates some of the risks and costs pollution entails. About 600,000 metric tons of oil enter the ocean every year from natural seeps, all of which the ocean has successfully assimilated through the ages. But as oil came to play an increasing role in human affairs, the volume of oil entering the ocean multiplied manyfold. Two-thirds of all the oil produced in the world is now shipped by sea. Although transportation practices have been improving over the years, these improvements have not kept pace with the growth in the volume of oil shipped. More than 6 million metric tons now flow into the seas annually, more than one-third of which comes from such routine tanker operations as spilling while loading and unloading, discharging ballast, and cleaning tanks. The floating lumps of tar that can be found on all the oceans and on many beaches bear witness to this calamitous trend.[3]

Less apparent, but in the long run perhaps more dangerous, are those portions of the petroleum that disappear into the sea. No one knows what all this oil will ultimately do to marine fisheries or to the complex ocean ecosystem. A UN report has noted that "the fact remains that once the recovery capacity of an environment is exceeded, deterioration can be rapid and catastrophic; and we do not know how much oil pollution the ocean can accept and still recover."[4] Yet many standard projections show the volume of ocean oil traffic expanding up to six times before world petroleum production peaks and begins to decline.

In addition to the general threat to the oceans, a more specific threat already plagues narrower stretches of water. Although tanker accidents account for less than 5 percent of all marine oil, a large spill concentrated in a single area can be more devastating than a multitude of smaller dispersed discharges. At the end of World War II, the world's largest oil tanker could carry about 18,000 tons. About a decade ago, a race of giant tankers emerged; the capacity of a single oil carrier grew to 100,000 tons and even 250,000 tons and larger. The *Globtic Tokyo* carries 483,664 metric tons—some 3.6 million barrels of oil. Requiring twenty minutes and three miles to stop, these unwieldy supertankers invite accidents, and several have broken up in heavy storms. As Eugene Coan of the Sierra Club observes, "If you have an accident with a very large ship, you're likely to have a very large accident."

Similar phenomena beleaguer other forms of energy growth. To be sure, increasingly stringent controls can be applied, but the costs of enforcing and complying with such controls eventually operate as a capital constraint. Pollution controls now commonly constitute more than one-third of the total cost of a new energy facility, and in many cases it is far from clear that such controls are adequate. Moreover, some kinds of pollution, such as carbon dioxide, simply cannot be controlled except by burning less fossil fuel.

Material Constraints

Scant attention has been paid to the material requirements of various energy technologies. While we now have a reasonably clear idea of the energy requirements of steel production, we have no similarly detailed accounting of the steel requirements of energy production. Yet various types of steel will be absolutely necessary for the construction of oil wells in the Middle East, pipelines across the Soviet Union, power plants in Europe, transmission facilities in Brazil, and virtually every other energy-related device.

Different energy technologies demand different materials. Gallium arsenide photovoltaic cells, used to generate electricity from sunlight, require gallium; ultra-efficient cryogenic electrical transmission systems need helium. The most efficient fuel cells yet developed use platinum as a catalyst; the amount of platinum that such cells would require annually if half of all U.S. electricity were produced with fuel cells would exceed the present yearly world production. Titanium may prove to be the limiting factor on ocean thermal electrical plants, and even copper production seems unlikely to keep pace with the extra demands of new energy technologies.[5]

Politics as well as general scarcity may lead to material shortages. Scattered unevenly through the earth's crust, some crucial minerals are concentrated in relatively few lands, many of them Third World nations. Such countries have for years been selling in a competitive market, but buying from what they perceive as multinational cartels. In the wake of the OPEC success, and in the midst of calls for a new international economic order, the mineral-rich nations may well decide to turn the tables.

Various material shortages may hinder energy growth in different ways. For example, although water is obviously in great global abundance, a lack of sufficient local water makes the construction of synthetic fuel facilities at otherwise suitable sites impossible.[6] Sometimes a lack of spare parts, of manufacturing capacity, or of transportation equipment will delay production temporarily. Coal production in the United States may be limited for the next ten years by a simple lack of railroad cars.

The most intractable limits are those posed by needs that no known materials can satisfy. The materials needed for the "first wall" of fusion reactors must be able to withstand conditions so extreme that no existing test facilities can simulate them.

Financial Constraints

Capital represents the "seed corn" of all economies, the capacity for sustained production. A society that eats its seed corn—in this case, by spending too much of its income on goods and services, and saving too little for investments in future production—has a bleak future. The argument over whether the world faces a capital crisis has generated almost enough heat to solve the energy crisis. The issue is complex, and contrary opinions are rooted in different assumptions about economic growth, government spending, inflation, business cycles, and a host of other variables.[7]

Capital, by its very nature, is limited. Within a finite capital budget, tough choices must be made. Such choices are usually evaluated in terms of cost per unit of productive capacity. One automobile plant, for example, is compared with another in terms of how much investment each requires per car per day. For energy investments, an analogous figure is the investment needed to produce—or to save—the energy equivalent of one barrel of oil per day. When the capital cost of producing one barrel of oil exceeds the capital cost of conserving it, the most productive investments will be those made to heighten efficiency.

From the end of World War II until quite recently, the capital cost of producing fuel remained low. For example, the investment needed (in wells and pipelines) to produce Middle Eastern oil at the rate of one barrel per day ranges from $50 to $250. Amortizing these investments over the lifetime of the field reduces the cost of oil to just a few cents per barrel. In contrast, oil from the North Sea is expected to require an investment of $10,000 per daily barrel; Arctic oil and gas will require between $10,000 and $25,000 per daily barrel; and synthetic fuels from coal will demand an investment of from $20,000 to $50,000 per daily barrel.[8] To obtain the thermal equivalent of a daily barrel in the form of electricity from a new power plant requires an investment in excess of $100,000.

The capital costs of fuel production, which include the costs of extraction and of combustion, increase greatly as higher environmental standards and tighter health and safety regulations are put into effect. Generally, however, this merely means that prices are being adjusted to "internalize" costs that were previously inflicted on society but were not explicitly accounted for. The higher prices reflect the cost of preventing black lung disease among coal miners or of decreasing the likelihood that a catastrophic accident will take place at a nuclear power plant.

The costs of oil, coal, and shale-derived oil can only rise. When the Alaskan oil pipeline was proposed in 1969, the estimated cost of the project was $900 million; before it was completed in 1977, total costs had soared to nearly $8 billion. The cost per ton of underground coal mining capacity has doubled over the last five years. Atlantic Richfield bowed out of an oil shale complex when its projected costs tripled in three years.

The electrical utility industry is the most capital-intensive of all industries—requiring, for example, four times as much investment per dollar of revenues as the steel industry.[9] And recent escalations in construction costs have dealt the industry a staggering blow. Construction costs for nuclear power plants have more than quadrupled in recent years. During the thirteen years that the Kaiparowits coal-fired plant in the American Southwest was under consideration, its projected size was cut in half while its projected costs soared sevenfold. A recent report to the U.S. Federal Power Commission concluded that a 6 percent electrical growth rate would require at least $650 billion for new facilities over the next fifteen years, compared with $145 billion over the last fifteen.

As long as conventional sources supply most of the energy the world uses, upward cost trends are here to stay. Fuels will not become more plentiful and accessible; on the contrary, the best deposits will be exhausted. And as the biosphere becomes more saturated with pollutants, even more rigorous and expensive environmental controls will have to be imposed.

It is sometimes argued that renewable energy sources will provide an escape from the rising costs associated with the depletion of finite resources. The sun is expected to provide the earth with a rather steady, free flow of energy for billions of years. However, such reasoning is a little too simplistic. Only a limited number of choice solar sites exist: areas with three hundred days a year of unclouded sunlight, with steady winds of 30 mph or more, or with large volumes of falling water. Most such sites lie far from the areas that currently demand energy, and as more remote sites are employed, costs will rise.

Renewable energy sources also tend to be expensive to tap. Just how much the new equipment will cost when it is manufactured by mature industries enjoying the economies of mass production is hard to say. But it is unlikely to be cheap. Today, photovoltaic cells are several times as expensive per peak watt as nuclear power plants. The cost of wind power appears to be roughly comparable to the cost of nuclear power. The expenses entailed by different bioconversion options vary, but most appear to be at least as costly as processes using coal.

Enormous sums of capital would be required to build enough new energy facilities to meet all projected demands. Two trillion dollars is considered by some to be a conservative estimate of the combined energy-related capital needs through 1985 of Europe, Japan, and the United States if conventional options are pursued. On the other hand, much of this capital could better be used to refashion our living environments, redesign our transportation systems, and reshape our industries to obviate the need for much of this energy. Because capital is limited, huge investments in energy supplies may be taking money away from far more productive investments in increased efficiency.[10]

Political Limits

Every unit of energy, regardless of its source, entails costs, and the true costs are often not borne by the beneficiaries. The losers in the trade-off have grown restive in recent years, and energy battles are now being fought in every corner of the political landscape. Nuclear power plants, strip mines, oil refineries, deep-water ports, hydroelectric facilities, and high-voltage power lines are both the issues and the plunder of a struggle that transcends traditional ideological boundaries.

The opposition is both private and public. Carolyn Anderson, a

Wyoming rancher whose land lies over a rich coal vein, draws the line clearly. "Don't underestimate us," she says. "We are descendants of those who fought for this land, and we are prepared to do it again." The governor of Colorado, a state rich in coal and oil shale, was elected on a platform that promised Coloradans that their state wouldn't "become the nation's slag heap."

Fuel use harms the environment more than any other human activity does; it scars the landscape, heats the atmosphere, generates tons of pollutants, and creates dangerous radioactive by-products. When energy is used for necessary purposes, some such costs can certainly be justified. But to increasing numbers of people, the costs of continued energy growth now seem to outweigh any perceptible benefits.

Opposition to the expansion of fuel facilities is most pronounced in the industrial countries. Building a centralized energy facility anywhere in Europe, Japan, or North America has become difficult indeed. Although a majority of the citizens in those regions would probably not ask for zero energy growth, very few want a new power plant in their neighborhood, and every possible site is in *somebody's* neighborhood.

In effect, the developed world has run out of space: geographical space, environmental space, and psychological space. Where once many activities could grow independently, now each one can grow only by impinging on the others. Illinois provides a telling case study of the competition among different kinds of spatial needs.[11] With more bituminous coal than any other state in the United States, Illinois also has much of the country's best agricultural land. But land cannot simultaneously be a strip mine and a cornfield, and the same water cannot be used by a coal gasification plant and by farmers to irrigate fields. Some evidence suggests that effluents from energy facilities may already be affecting the state's agricultural production negatively; with continued growth, production shortfalls are an eventual certainty. Illinois agriculture is as energy-intensive as any farming system in the world, and farmers have traditionally favored energy growth. But many have now begun to draw the line, fighting strip mines, dams, nuclear power plants, and any other developments that will take additional fertile land out of production.

While energy forecasters plot their demand curves toward infinity, people throughout much of the industrialized world are demanding an end to open-ended growth. Few would phrase it like that. They do not oppose the use of gasoline; they just oppose this particular refinery. They do not oppose nuclear power; they merely feel that this particular reactor is poorly sited and unnecessary. But when such attitudes are widespread, *every* refinery and *every* reactor will be opposed. Whereas civic boosters used to talk of luring new power plants to an area to "capture the benefits of growth," they now increasingly must beseech residents to "responsibly shoulder the burdens of growth." But most people are less enthusiastic about shouldering burdens than about receiving benefits. The resulting political self-adjustment, which includes weighing total costs against total benefits and rejecting further growth, may well prove to be among the most important limiting factors in energy development.

The Coming Energy Transition

During the last twenty-five years, world fuel consumption tripled, oil and gas consumption quintupled, and electricity use grew almost seven-fold.[12] Clearly, such trends cannot be sustained indefinitely—nature abhors exponential curves as well as vacuums.

The world has begun another great energy transition. In the past, such transformations have always produced far-reaching social change. For example, the substitution of coal for wood and wind in Europe accelerated and refashioned the industrial revolution. Later, the shift to petroleum altered the nature of travel, shrinking the planet and completely restructuring its cities. The coming energy transition can be counted upon to reshape tomorrow's world. Moreover, the quantity of energy available may, in the long run, prove much less important than where and how this energy is obtained.

Most energy policy analyses do not encompass the social consequences of energy choices. Most energy decisions are based instead on the naïve assumption that competing sources are neutral and interchangeable. As defined by most energy experts, the task at hand is simply to obtain enough energy to meet the projected demands at as low a cost as possible. Choices generally swing on small differences in the marginal costs of competing potential sources.

But energy sources are *not* neutral and interchangeable. Some energy sources are necessarily centralized; others are necessarily dispersed.

Some are exceedingly vulnerable; others will reduce the number of people employed. Some will tend to diminish the gap between rich and poor; others will accentuate it. Some inherently dangerous sources can be permitted unchecked growth only under totalitarian regimes; others can lead to nothing more dangerous than a leaky roof. Some sources can be comprehended only by the world's most elite technicians; others can be assembled in remote villages using local labor and indigenous materials. In the long run, such considerations are likely to prove more important than the financial criteria that dominate and limit current energy thinking.

Appropriate energy sources are necessary, though not sufficient, for the realization of important social and political goals. Inappropriate energy sources could make attaining such goals impossible. Decisions made today about energy sources will, to a far greater extent than is commonly realized, determine how the world will look a few decades hence. Although energy policy has been dominated by the thinking of economists and scientists, the most important consequences may be political.

After consideration is paid to the myriad constraints facing energy growth, and to the sweeping social consequences produced by energy choices, few attractive options remain. For reasons that will be elaborated in chapters 2 and 3, the long-term roles of fossil fuels and nuclear fission are likely to be modest. Geothermal power is already proving useful in Italy, Iceland, New Zealand, and the United States as a means of generating electricity and as a source of space heating. However, the exploitable global geothermal potential appears to be rather small, and the environmental impact of geothermal operations is larger than most people assume.[13]

Nuclear fusion is popularly envisioned as a clean source of virtually limitless power. But the reality belies the ideal.[14] William Metz has noted "a gap . . . between what the fusion program appears to promise and what [it] is most likely to deliver." While some advanced fusion cycles—most notably those that would fuse two deuterium nuclei or that would fuse a proton with a boron nucleus—could theoretically provide a nearly inexhaustible source of relatively clean power, such reactions are *very* much more difficult to achieve than the deuterium-tritium reaction that is the focus of almost all current research. For example, the hydrogen-boron reaction requires temperatures of 3 billion degrees Centigrade, whereas the deuterium-tritium reaction can take place at 100 million degrees. When scientists speak of building a commercial nuclear fusion reactor within twenty-five years, they are referring to a deuterium-tritium reactor, a reactor that does not share all the idealized characteristics associated with nuclear fusion. The D-T reactor's fuel supply would not be limitless; tritium is derived from lithium, an element not much more abundant than uranium. The D-T fusion power plant might well be even larger (and hence more centralized) than current conventional facilities, and the energy produced could be much more expensive than that derived from current sources. The reactor would certainly require maintenance, but the intense radioactivity of the equipment would make maintenance almost impossible. Although cleaner than nuclear fission, a large fusion reactor might nonetheless produce as much as 250 tons of radioactive waste annually.

Even though a deuterium-tritium fusion reactor would be much "easier" to build than a device employing a more advanced fuel cycle, the pursuit of D-T fusion still represents the most ambitious engineering undertaking in human history. Current experimental fusion devices are enormous energy "sinks" that consume far more energy than they produce. Because of the exceptional difficulties involved in achieving a net energy gain from fusion, the first generation of fusion reactors may not be designed to optimize power production. Rather, they may be hybrid fusion-fission devices designed to convert non-fissionable uranium into plutonium fuel for fission reactors. This hybrid technology, now being pursued by the Soviet Union and under active consideration in the United States, would combine the most unattractive features of nuclear fission with the incredible complexities of nuclear fusion. It would be tragic if the resulting mix were marketed as "safe, clean nuclear fusion."

Renewable energy sources—wind, water, biomass, and direct sunlight—hold substantial advantages over the alternatives. They add no heat to the global environment and produce no radioactive or weapons-grade materials. The carbon dioxide emitted by biomass systems in equilibrium will make no net contribution to atmospheric concentrations, since green plants will capture CO_2 at the same rate that it is being produced. Renewable energy sources can provide energy as heat, liquid or gaseous fuels, or electricity. And they lend themselves well to production and use in decentralized, autonomous facilities. However, such sources are not the indefatigable genies sought by advocates of limitless energy growth. While renewable sources do expand the limits to energy growth, especially the physical limits, the fact that energy development has a ceiling cannot ultimately be denied.

The highest energy priority in all lands today should be conservation. Investments in saving energy, whether to double the efficiency of an Indian villager's cookstove or to eliminate energy waste in a steel mill, will often save far more energy than similar investments in new power facilities can produce. The cheapest and best energy option for the entire world today is to harness the major portion of all commercial energy that is currently being wasted.

A transition to an efficient, sustainable energy system is both technically possible and socially desirable. But 150 countries of widely different physical and social circumstances are unlikely to undergo such a transition smoothly and painlessly. Every potential energy source will be championed by vested interests and fought by diehard opponents. Bureaucratic inertia, political timidity, conflicting corporate designs, and the simple, understandable reluctance of people to face up to far-reaching change will all discourage a transition from taking place spontaneously. Even when clear goals are widely shared, they are not easily pursued. Policies tend to provoke opposition; unanticipated side effects almost always occur.

If the path is not easy, it is nonetheless the only road worth taking. For twenty years, global energy policy has been headed down a blind alley. It is not too late to retrace our steps before we collide with inevitable boundaries. But the longer we wait, the more tumultuous the eventual turnaround will be.

I
Fading Dreams

2. The Future of Fossil Fuels

WHEN COLONEL E. L. Drake set up a drilling rig in 1859, near Titusville, Pennsylvania, the townspeople thought him unbalanced. Others before him had struck oil while drilling for water, but Drake was consciously *seeking* the nearly useless muck. Oil could only be peddled as a medical cure-all or burned in kerosene lamps, and most folks at that time preferred whale-oil lamps.

Drake's pioneering oil well proved successful. Not long after his strike, the American Civil War choked the nation's supply of whale oil, and history began to saunter unsuspectingly toward the petroleum era. The kerosene business evolved into the oil industry, which eventually produced a dozen petroleum-based fuels and thousands of petrochemicals.

Children of the petroleum era tend to forget how brief this period has been. Just fifty years ago, 80 percent of the world's commercial energy came from coal and a mere 16 percent from oil and gas. Even as recently as 1950, coal still provided 60 percent of the world's commercial fuel. For the next two decades, oil and gas consumption grew rapidly, passing coal use in 1960. Today oil and gas comprise two-thirds of the world's commercial energy budget.[1]

Oil and gas, like all other fossil fuels, are in finite supply. The actual size of the supply, and its likely rate of depletion, have become matters of controversy. Making a case study of the United States, where these issues first arose, is one way to gain insight into this continuing controversy.

The American Experience

The oil industry correctly advertises that "America runs on oil." But what they do not broadcast is that any country that "runs on oil" must eventually run out of it. Nineteenth-century oil producers were aware of the limits of their known resource base, but with the 1901 Spindletop gusher in southeast Texas, heady success overpowered prudence. The inevitability of oil exhaustion became an abstraction—hard to grasp and comfortably remote—as huge discoveries were made in Oklahoma, Louisiana, California, and Alaska. In recent decades, cheap, plentiful oil has been substituted for capital, for labor, and for other materials, influencing the shape and behavior of modern America as no other commodity has. As more and more oil was pumped into the veins of American manufacturing, commerce, and transportation, the oil industry came into unprecedented economic and political power.[2]

At mid-century, few critics were ready to challenge the oil companies. But in early 1956 a blow was dealt from within. M. King Hubbert, a geologist with Shell Oil, was then at work on an address to a conference sponsored by the American Petroleum Institute. Worried about the exponential increase in the rate of U.S. petroleum consumption, Hubbert resolved to use his speech at the oil industry's forum to make public his concern.

In 1956, the ultimate recoverable petroleum resource base of the United States was commonly pegged at about 150 billion barrels. Since the nation had consumed only 50 billion barrels of oil during the industry's hundred years of operation, an ultimate resource base three times that large was generally believed to afford the country a comfortable margin of time in which to find petroleum substitutes.

But Hubbert demonstrated that geological exploitation follows a predictable pattern, that "in the production of any resource of fixed magnitude, the production rate must begin at zero, and then after passing one or several maxima, it must decline again to zero."

In his key illustration, Hubbert drew a production curve for petroleum on a grid, with each rectangle representing 25 billion barrels of oil. The curve representing all U.S. oil production—yesterday and forever —could cover only six rectangles, or 150 billion barrels. As of 1956, the oil represented by two rectangles was already spent. When three rectangles were covered, half the oil would be gone and production would begin to decline. Hubbert calculated that the third rectangle would be covered within ten years. If the U.S. oil resource base turned out to be 200 billion barrels instead of 150 billion—an increase equal to the total content of eight oil fields the size of the mammoth east Texas find—the halfway point in production would be delayed only five years. In essence, Hubbert demonstrated that U.S. oil production would "peak out" in ten to fifteen years, and then begin a slow, steady decline back to zero.[3]

When executives at Shell read over Hubbert's prepared remarks, they were understandably horrified. Minutes before his San Antonio presentation, Hubbert received a telephone call from headquarters asking him to delete the "sensational" portion of the address. He refused, and the great American oil controversy began.

Hubbert's chart caught everyone off guard, but no one effectively challenged its logic. Although the shape of the curve could be altered somewhat by changes in consumption rates, the *fact* of the curve would remain inviolate. Retarding the consumption growth rate would postpone the date at which maximum oil production was attained, but not by more than a few years. Moreover, no one, least of all the oil industry, was ready to crimp the oil consumption growth rate in 1956.

However, the day of reckoning could be put off. If the total area under the curve, the estimated oil resource base, were found to be larger than was commonly believed, the apex of the depletion curve would be shifted rightward on the time axis accordingly. Predictably, every major oil-related institution in the United States began re-examining its estimates of the nation's petroleum resource base.

To understand the figures that the petroleum industry came up with, it is necessary to understand the difference between resources and reserves. "Reserves" are deposits of minerals in known locations that can be recovered profitably with existing technology. They represent the industry's immediate working stock, and are not an index of the total resource base. Without this understanding, a person looking at U.S. oil reserves over time would have to conclude that oil is being manufactured in the earth's bowels. No matter how much fuel is consumed, we always seem to have ten more years' supply in reserves. In fact, new discoveries, technological advances, and rising prices simply put more resources into the "reserves" category. "Resources," on the other hand, include not only all reserves, but also all fuel that is known to exist but that cannot be recovered at current prices and with current technology, and an estimate of fuel deposits as yet undiscovered.[4]

Even the concept of "recovery," as energy students soon discover, may also require a word of explanation. Oil fields are popularly misconceived as underground lakes of fluid petroleum. Actually, oil fields are oil-soaked sand and rock, generally harder and less porous than set concrete. Bringing the oil to the surface is not a simple matter of inserting a straw and sucking. Commonly, about 10 percent of the oil in a field can be forced to the surface by reservoir pressure. Another 20 to 25 percent can be pumped up. Additional oil can be extracted only by using secondary and tertiary recovery practices—heating the area, and flooding it with fluids and chemicals. Complicated, expensive, and energy-intensive, such practices have so far been less than successful.[5]

Estimating undiscovered resources is necessarily a speculative enterprise. Oil resources are particularly hard to gauge, for oil can only be found with a drill. Until the bit actually strikes oil, all is guesswork. However, the guesswork has grown impressively sophisticated. In the industry's early days, wildcatters depended primarily upon oil seeps to track down reservoirs. Later, prominent geological formations were "read" to locate undiscovered oil. Today, the gravity meter, the airborne magnetometer, and the reflection and refraction seismograph are the tools of the search. However, most clues still lead into blind alleys. One hundred new-field wildcat wells are sunk in the United States today for each new field of one million barrels or more discovered—yet one million barrels will sustain the United States for only ninety minutes.

Oil prospecting remains detective work largely because "strike" conditions can vary so wildly. Oil deposits have been found within one hundred yards of the surface and more than three miles beneath it. Reservoir widths range from a few hundred yards to more than a hundred miles. When an oil deposit is buried far underground, and especially when the ground itself lies beneath a quarter mile of seawater, examining the resource to establish its volume and quality poses obvious difficulties. Thus, even after a reservoir is discovered, years of uncertainty often intervene before its true extent is "proved."

The whole field of petroleum resources estimation is charged with controversy. Competent, well-intending authorities, armed with different assumptions and methodologies, splash their numbers all over the board. In the furor that followed Hubbert's 1956 speech, a rash of higher estimates of the petroleum resource base appeared. Claims that oil resources amounted to 204, 250, 372, 400, and even 590 billion barrels were made over the next few years.

To the outsider, a total lack of agreement among the experts in their estimates strongly suggests that the experts don't know what they are doing. Or at least that some of them don't. In fact, no one "knows" how much oil is down there, or where it is. Estimates of undiscovered resources depend upon inferences from objective information: mountainous piles of data on geological formations, seismic tests, total number of wells attempted, total feet drilled, volumes of oil discovered, and so on. Creative forecasting, which consists of putting key variables together in ways that lend insight into how much oil remains to be discovered, involves great inductive leaps.

Most of the evidence accumulated in recent years appears to support Hubbert: U.S. oil production did peak in 1970, as Hubbert had predicted fifteen years earlier, and began a steady decline. A Geological Survey study issued in May of 1975 indicates that the undiscovered U.S. oil resource base lies within the range of 50 to 130 billion barrels, with a 95 percent probability at the lower figure and a 5 percent probability at the higher one. A National Academy of Sciences report released earlier that year reached similar conclusions.[6]

Regrettably, America's oil is now almost certainly half gone. The optimists who expected oil production to increase for so many decades that there was no need to worry about the eventual decline are now few in number. Instead, most oil watchers currently believe that the 1970 production peak in the forty-eight contiguous states was indeed a one-time peak. The present clash of views centers largely upon how rapidly the United States will slide down the far slope of the oil depletion curve. The downhill pace will be determined by the extent of the Alaskan resources, the quantities of oil obtained from the continental shelves, and the rate at which advanced oil recovery technologies are developed and implemented. The authors of the Project Independence report in 1974 thought that these three factors could lead to a brief production increase by 1985. Indeed, their prophecy might even be fulfilled if enough money is poured into the single-minded goal of increasing the rate of oil extraction. But such a policy would provide precious little energy per dollar of investment and would only make the post-1985 decline that much more precipitate.

The total quantity of undiscovered oil will not be known until it has all been discovered. But nobody is down there brewing more oil. And the more that is learned about the size of the ultimate U.S. oil resource base, the smaller that base appears to be.

The United States houses most of the international oil industry, as well as many of the world's most distinguished schools of petroleum geology. No other large land mass has been as extensively probed as has the United States, where oil—together with natural gas—comprises fully three-fourths of all commercial fuel used. With about 10 percent of the world's potential oil-bearing areas, the United States has a drilling density about seven times higher than the world average. Thus, examining the U.S. experience can provide a basis of comparison for analyzing world oil resources. Even rough agreement on the extent of the remaining U.S. oil supply was not achieved until a full five years after oil production had peaked. Yet, compared with what is known about U.S. oil deposits, information about the fossil fuels in the rest of the world is downright sketchy.

World Oil and Gas

Most estimates of the world petroleum resources, like the U.S. estimates discussed above, are based on a combination of historical discovery patterns and geological analogies. The score of published estimates, and additional unpublished estimates that have been produced since 1950, mostly range between 1.2 trillion barrels and 2.5 trillion barrels. Most relatively recent estimates have tended to cluster between 1.8 trillion and 2.0 trillion barrels.[7]

Though disagreements arise over the ultimate volume of recoverable oil, a general consensus exists about how the oil is distributed. The Middle East has roughly 30 percent of the world's oil, of which one-tenth has been consumed. The Soviet Union has about 25 percent, of which one-twelfth has been consumed. The United States and Africa each have about 10 percent; one-half of the U.S. oil has already been consumed, while all but one-twentieth of Africa's remains in the ground. Latin America is generally believed to have 8 percent of the world total, of which about one-fifth has been consumed.[8] Western Europe, including the North Sea, has less than 4 percent of the expected world total, of which an almost negligible amount has been consumed. (The enormous attention focused on North Sea oil is more a consequence of the resource's location than of its size.)

The United States, Western Europe, and Japan face an immediate oil squeeze. Most other areas have ample oil to meet their domestic requirements for some time yet. But the oil-short areas encompass most of the world's industrial base, and they all expect to import prodigious amounts of oil from the oil-rich regions.

In 1973, the growth of petroleum consumption was interrupted by the Arab boycott. Such growth is unlikely to resume. A fivefold increase in oil prices has already cut deeply into the growth rate, and further price increases are certain.

Oil price rises have political causes and economic effects. Much of the remaining supply of easily obtained oil is in single-resource nations that intend to stretch their income from this source of wealth as long as possible. Moreover, at least some oil-producing countries understand that oil has more value as a petrochemical feedstock than as a fuel, and these countries can be counted upon to save as much of their petroleum as possible for non-energy purposes.[9] With effective monopoly control held by a few major producing countries, global oil use will probably not be allowed to grow exponentially to 1990—when, if past rates of production increase were to continue, world oil production would probably peak—and then plummet as more and more wells run dry. World oil output is more likely to rise for three or four more years, and then to stabilize at that level for several decades. The Middle East might temporarily slow down production to buffer any brief surges (of rather high-priced oil) from the North Sea and elsewhere.

The problems of estimating recoverable oil resources reflect the difficulties surrounding the extraction of oil from reservoirs. Natural gas exhibits no such problems. Once tapped, it surfaces. Gas estimates do, however, entail many other problems.

A fixed quantifiable relationship between gas and oil is presumed to

exist, and gas resource estimates are generally derived from oil resource
estimates. But the historic gas/oil ratio may be changing. Further, more
gas fields that are unassociated with oil fields are now being discovered,
and, as drilling rigs capable of probing deeper and deeper have revealed,
the ratio of gas to oil seems to increase at the lower depths.

Even though the magnitude of total oil resources remains in question, gas resources are predicted by using this controversial estimate in
conjunction with a dubious gas/oil ratio. The resulting estimates obviously vary tremendously. The world total for natural gas is commonly
hypothesized to be about 12 quadrillion cubic feet, although the most
recent authoritative study—done for the Ninth World Petroleum Conference—claims that the resource base may be only half this large.[10]
(Current world consumption of natural gas is about 15 trillion cubic feet
a year.)

Another much smaller source of fuels and petrochemical feedstocks
is to be found in the natural gas liquids. If presumed ratios of natural
gas liquids to natural gas (the reader no doubt recognizes that we are
beginning to presume ourselves uncomfortably far out on a limb) prove
to be accurate, the world resource base totals about 400 billion barrels
of natural gas liquids, or roughly 2 percent of the estimated volume of
oil.

Coal: A Transitional Fuel

Coal, the world's most plentiful fossil fuel, has been used for at least
two thousand years. The Chinese burned coal, and evidence suggests
that the classical Romans did as well. Coal consumption increased
steadily in Europe from the fourteenth century on, as the brick, glass,
and iron industries became coal burners. By the mid-sixteenth century,
England was mining about 200,000 metric tons of coal a year, and with
the advent of the industrial revolution in the eighteenth century, coal-
fired operations increased dramatically. By 1925, the world was producing 1.3 billion metric tons of coal a year. By 1975, the figure reached
3.25 billion metric tons, of which Europe accounted for about 36 percent, the Soviet Union for about 23 percent, the United States for
approximately 17 percent, and the People's Republic of China for about
14 percent.

Because solids are easier to measure than liquids or gases, coal resource estimates are probably more reliable than oil and gas estimates. Total world coal resources most likely amount to between 7 and 10 trillion metric tons. If all that coal were potentially available—which it certainly is not—the world fuel resource base would be bountiful. Even if our current rapid rate of growth continued, coal extraction could not peak until some time after 2200 A.D. Annual production would then be about 24 billion tons a year—eight times higher than the present output.

All this coal will never be mined, however. Much of it rests in beds too thin or too deep to be mined. Moreover, at some point more energy is used to extract the last bit of coal from deep in the earth than the coal itself contains. Long before this point is reached, the economics of coal production will prove impossible.

A reasonable estimate of the recoverable coal resources—yet one that still takes major advances in extraction technology and substantial price increases into consideration—is about 2 trillion metric tons. This amount of coal could support the world's current level of coal use for almost a thousand years, or it could sustain current world levels of consumption of all fossil fuels for over two hundred more years. But environmental alarms are likely to halt coal combustion long before then. In particular, the buildup of carbon dioxide in the atmosphere will almost certainly prove intolerable long before all the world's recoverable coal is consumed.

Although no worldwide coal shortage threatens, some geographical areas are in comparatively poor shape. Europe faces what could be a coal crisis. European coal extraction, for example, now constitutes 36 percent of the world total, but Europe has only 6 percent of the world's remaining coal. In contrast, both Latin America and Africa face another kind of resource pinch. Together, they have less than 1 percent of the world's total coal. Since both areas have low coal consumption rates, their present problem is that of limited potential rather than of impending crisis.

Three countries contain more than 80 percent of the world's estimated coal supply. The Soviet Union's share, 56 percent, is enormous, while the United States owns a hefty 19 percent and the People's Republic of China has about 8 percent. As production of other fossil fuels peaks and declines, this skewed distribution of coal may prove politically significant. The Soviet Union, for example, has a much higher percentage of the world's coal than the Middle East has of the world's oil.

Coal, despite its geographical concentration, is a relatively bountiful fuel. Because it is less valuable than petroleum as a chemical feedstock, over the long term the substitution of coal for fuel oil where possible makes sense. Coal is likely to play a prominent role in the coming energy transition, and this role will expand to the degree that the projected expansion of nuclear power is successfully halted. But coal should be viewed strictly as a transition fuel. Over and beyond ultimate resource constraints and the climatic alterations associated with increased amounts of atmospheric CO_2, coal holds but faint attraction as a long-term energy option.

Formidable environmental problems attend both the extraction and the combustion of coal. Underground mines may cause surface lands to subside; they may entail drainage problems (11,000 miles of American streams are afflicted); and they pose serious threats to the health and safety of the miner who faces slow death by black lung disease and quick death in a cave-in. While all these dangers can be mitigated, none can be eliminated.

A 1,000-megawatt power plant annually consumes the production from twenty miles of a surface mine with a 225-foot wide bench and a three-foot coal seam. The reclamation of land sacked by surface mines has in many parts of the world been the exception rather than the rule. Capitalizing their profits while socializing their costs, American coal companies left behind 20,000 miles of unreclaimed strip mine benches in Appalachia alone. The Germans, on the other hand, have an outstanding record of strip mine reclamation. They are even reclaiming the "world's biggest hole"—a four-square-mile 1,000-foot deep lignite mine near Bergheim that is moving north at a relentless three feet a day. But even reclaimed land, while ransomed from aesthetic oblivion, is often worth less in its rejuvenated than in its virgin state. Under ideal conditions, reclaimed land can often support only pastures, not more valuable row crops. In arid and semi-arid regions, reclamation of any sort is nearly impossible.[11]

Coal combustion produces emissions of fly ash, sulfur oxides, toxic metals, and carcinogenic organic compounds. It entails the release of more mercury than any other human activity does. Precipitators can remove up to 99 percent of all ash, but can catch only half the minuscule ash particulates that are most injurious to human health. Lead, cadmium, antimony, selenium, nickel, vanadium, zinc, cobalt, bromine, manganese, sulfate, and certain organic compounds cling to these small particulates, against which evolution has provided the human respiratory system with no satisfactory defense.[12]

Considerable evidence now suggests that the sulfur in coal is most troublesome in two forms: as sulfuric acid or as sulfate salts. Acid rains have been a recognized problem for decades in Scandinavia, where they kill fish and reduce agricultural and timber harvests; similar rains now fall in many other parts of the world. The only entirely effective sulfur control program to date has entailed a switch to low-sulfur coal—an approach with obvious long-term limitations. Other approaches have included erecting tall stacks (some approaching the Empire State Building in height) to dilute the pollutants, using intermittent controls (reducing or even halting combustion when atmospheric conditions are poor), installing scrubbers (to physically remove sulfur from flue gases), and employing a variety of other techniques to remove sulfur from the coal before or during combustion.

Tall stacks and intermittent controls do not provide a long-term answer to a growing problem; if more power plants are built, concentrations will again reach hazardous levels. Moreover, tall stacks may, counterproductively, enable sulfur dioxide to remain airborne longer, increasing the likelihood that some percentage of it will oxidize into sulfuric acid. Scrubbers are expensive and energy-intensive, and have been riddled with technical difficulties. Most current scrubbers produce 8 or 9 cubic feet of sludge per ton of coal burned, so a 1,000-megawatt plant fueled by high-sulfur coal would have to dispose of 80,000 cubic feet a day. A new power plant in Pennsylvania plans to fill a five-mile stretch of valley four hundred feet deep with sludge over the next twenty-five years. The long-term ability of such sludge deposits to withstand, for example, bacterial attacks that could release hydrogen sulfide gas is unproven. Regenerative scrubbers, which produce sulfuric acid or elemental sulfur and reuse their scrubbing agent, are under development, but these are expected to cost much more than the variety now in use.

Over the long term, removing sulfur from coal before or during combustion probably makes most sense. Since sulfur is more easily removed from gases and liquids than from solids, much coal research has centered on gasification and liquefaction. Some such technologies produce products that can be substituted for natural gas and petroleum, albeit at far higher prices. More than 150 companies around the world manufactured coal gasification equipment in the 1920s, and wartime Germany ran much of its economy on synthetic fuels derived from coal. The world's only working plant that produces liquid fuels from coal is located in South Africa; this refinery produces a variety of products, including gasoline, but does so rather inefficiently. Energy inefficiency, in fact, plagues all processes for deriving synthetic fuels from coal; net losses of from one-third to one-half of all the energy originally in the coal are sustained during the conversion. Although the sulfur content of synthetic fuels can be reduced to acceptable levels, fuel conversion plants are exorbitantly expensive to build, often require enormous quantities of water, and are ensnared in environmental problems of their own.[13]

Fluidized-bed combustion appears to be the most attractive coal technology at this time, though it probably doesn't deserve the unqualified praise sometimes heaped upon it. In a fluidized bed, air flows up through the boiler, suspending a hot bed of coal and limestone. Because its efficient heat transfer allows it to operate at relatively low temperatures, the fluidized-bed process does not produce the melted ash and nitrogen oxides that plague other coal technologies. However, the current generation of fluidized beds removes only about 90 percent of the sulfur in coal, and removes it in the form of calcium sulfate, which itself poses a disposal problem. In addition, the extent to which the use of large fluidized beds will control particulates is not known. Fluidized-bed technology should be relatively cheap, compact, and efficient when compared to conventional boilers with scrubbers.

A 30-megawatt fluidized-bed boiler began operations in West Virginia in late 1976, and the Tennessee Valley Authority has announced plans to build a 200-megawatt unit. Many smaller commercial models have been operated successfully in Europe, and have proven effective for use in small-scale decentralized electrical generation and in the district-heating of buildings. Our knowledge of the potential of this promising technology for large-scale application should expand in the next few years.

The clean coal-combustion technologies should be temporarily embraced by societies with ample coal. But exotic new coal technologies should not lure vast sums of money away from investments in sustainable energy sources that hold far more appeal over the long term. Coal should be viewed as an interim fuel, to be used efficiently to smooth the transition from the petroleum era to the solar age.

Oil Shale and Tar Sands

Bituminous sands, also known as heavy-oil sands or tar sands, contain a heavy, viscous raw oil mixed with grit. The geological origins of bituminous sands are disputed, but the oil they contain is from the same chemical family as petroleum.

Deposits of bituminous sands have been found in ten countries on all continents. The largest and best-mapped deposits are in northern Alberta, in Canada, although preliminary evidence suggests that Colombia also has large deposits. The Athabasca and other Canadian deposits are thought to contain about 100 billion barrels of recoverable synthetic crude oil. Oil recovery from bituminous sands has been attempted in the Soviet Union, Romania, Albania, and Trinidad. Great Canadian Oil Sand, Ltd., has been in operation since 1966, using open-pit mining. None of these efforts, however, has turned a significant profit.

Oil shale was formed in large, shallow, semi-stagnant bodies of water. The hydrocarbon content appears to be derived from algae, pollens, and waxy spores and takes the form of a solid known as kerogen. It differs markedly from petroleum in chemical composition, and poses special refining problems. The chemical energy bound in the earth's oil shale deposits is enormous—perhaps equal to 5 trillion barrels of oil. However, such gross figures mean nothing. While high-grade shale may yield more than 100 gallons of oil per metric ton, poorer grades may contain almost no recoverable oil.

In the Soviet Union and in China, some oil shale is crushed and burned directly under boilers. Limited quantities of synthetic oil have been produced from shale in Scotland and Estonia since the mid-nineteenth century, and related research efforts have long been under way in the United States, Brazil, and other countries. Yet no large-scale, commercially viable processes have yet been developed. Shale mining and refining pose formidable environmental problems, and require enormous amounts of water and energy. Much shale lies in dry areas, remote from energy markets. And because of its relatively low fuel content, a sizable fraction of the world's shale could probably be mined and processed only at a net energy loss. Most estimates of the recoverable oil shale range below 200 billion barrels. Although more oil shale may be obtained over time, especially as a feedstock for petrochemicals, economic and environmental factors will limit the amount produced in any one year to small quantities.

World Fossil Fuel Resources in Perspective

The recoverable energy in the world's fossil fuels is probably on the order of 10^{23} joules, of which 68 percent is found in coal, 30 percent in petroleum and natural gas, and 2 percent in oil shale and bituminous sands. Fossil fuels are currently being consumed at the approximate rate of 2.8×10^{20} joules per year. Thus, were fossil fuel use to continue at its current level, the world's resource base would not be exhausted for more than three hundred years. However, such lump-sum figures are misleading.[14]

First, fossil fuel consumption simply will not level off at the current rate; growth seems certain, for a while at least. Only one country, Sweden, has officially even looked into the probable consequences of zero energy growth. A few other countries have halfheartedly examined the possibility of reducing their energy growth rates modestly. Even if the industrialized countries were, voluntarily or forcibly, to opt for zero energy growth, the less developed nations could hardly be expected to follow suit.

Second, fossil fuel resources are unevenly distributed. About a third of the world's oil is in the Middle East. More than 45 percent of *all* fossil fuels are located in the Soviet Union. This Soviet hegemony may, in the sweep of history, far overshadow the current market disturbances caused by OPEC. In any case, the most vulnerable fuel "have-nots" will be Europe and Japan.

Third, all fuels are not created equal. Some are easily accessible; others are buried in the arctic tundra. Some can be cheaply transported and stored; others require much more costly handling. Some are exceptionally clean; others are dreadfully dirty. Such flagrant differences among fuels naturally determine the uses to which various fuels are put. For use in producing aviation fuel, for example, a low grade of Siberian coal with a high sulfur content ranks as a last resort at best. Yet oil and gas—choice fuels—are being consumed rapaciously and often unnecessarily.

Fourth, the most sensible fuel conservation strategy does not involve burning all fuels. Instead, much of this wealth should be husbanded for use as chemical feedstocks. Although many oil-based chemicals can theoretically be synthesized from materials other than petroleum—even from elemental carbon and hydrogen—such alternatives entail great expense and enormous energy investments.[15] For example, far more energy is required to assemble a petroleum molecule than is released when that molecule is burned.

A final qualification must be placed on these fuel estimates: they refer to gross energy stored in a fuel deposit and *not* to net energy available to perform work. In recent years, concern has mounted among energy analysts over the increasing energy investment required to produce, process, and deliver valuable fuels. Energy is needed to open mines and wells, to build and operate power plants and refineries, and to transport fuels and electricity from remote locations to major markets. This energy investment must be subtracted from the gross energy in unmined fuel to yield net available energy, the only energy that counts.[16]

The most accessible energy sources were tapped first, and increasing energy investments will be required to obtain the remaining fuel. Oil drilling goes deeper into the ground and farther into the oceans each year. Secondary and tertiary oil recovery techniques require prodigious energy investments. Only a small fraction of coal can be strip-mined; deep mines require larger energy investments to extract a smaller fraction of the coal in a deposit. When coal or oil shale is converted into synthetic oil or gas, a major part of its gross energy is lost during the metamorphosis.

Energy analysis, the new discipline that wrestles with net energy issues, provokes considerable controversy; the field is just beginning to attract the serious attention it deserves. One such controversy stems from different energy accounting practices various energy analysts use.

All analysts agree that the amount of energy used and lost at a coal conversion facility should be subtracted from the gross energy total. Most would further concur that the energy invested in building the facility—in refining its metals, and fashioning them into the end structure—should also be subtracted. But some argue that the energy needed to train and support the plant's workers, and even their families, should be subtracted as well. The drawing of such boundary lines is largely judgmental, though several conventions have been proposed.

A more difficult problem arises from the fact that energy has a qualitative as well as a quantitative dimension. Two-thirds of the energy in coal is lost in the process of producing electricity, but that electricity can provide more and better illumination than can a simple lump of burning coal. A warm lake contains far more energy than a small battery, but it is difficult to power a pocket calculator with a warm lake. Such realizations have led energy analysts to consider enthalpy and entropy, the qualitative dimensions of energy, as they make their calculations.

Historically, fuel consumption rises have followed doggedly on the heels of new discoveries, though with a lag time between discovery and use. Each year more fuel is discovered than the previous year; after a lag time, the consumption rate catches up. This pattern of rapid growth pushes all mineral exploitation into the bell-shaped curve that Hubbert plotted for U.S. oil extraction two decades ago. As long as production increases regularly every year, those extracting the resource become accustomed to growth and base their future plans upon expanding mineral wealth. But when production peaks and then begins to taper off, a society can be thrown into turmoil. If the decline is utterly unexpected, the consequences can be ruinous.

In 1492, the monarchs of Spain financed the explorations of Christopher Columbus to the New World. In the following century, mineral wealth from these newly found lands catapulted Spain to the height of its glory. Beginning in the 1520s, the flow of precious metals to the Iberian Peninsula grew more or less regularly for seventy-five years, making Spain one of the dominant states of Europe.[17]

In 1598, King Philip II died after a reign of forty years. Although the nation had a heavy burden of debt, resulting from stalemated wars with England and Holland, the debt was not onerous in the face of

Spain's rapidly increasing prosperity. When Philip III assumed the throne, Spain's prospects seemed bright. Unbeknownst to the Spanish rulers, however, the flow of gold and silver had already peaked: the next seventy-five years were years of rather steady falloff in production. However, traditional Spanish agriculture and small industry had languished during the nation's years of aggressive ascendancy, and were not successfully restored. The flow of precious metals had given Spain a golden moment in the sun, but the unanticipated decline in looted treasures brought the country to its knees.

The Spanish experience may hold special meaning for the contemporary world. The industrial nations have been shaped by the availability of cheap, plentiful oil at least as much as Spain was by the flow of gold. Unlike Spain, we can see the end ahead, and can choose to begin a voluntary transition, but failure to do so will lead to a fate much like Spain's.

The influence of our actions upon the future fossil fuel consumption curve is a weighty issue, an issue involving the value we attach to our progeny. Some fuel should certainly be saved for the future. But how much? We are certain to run out someday. Should we consume at a rate that will allow us to continue for fifty years? A hundred years? Five hundred years?

Economists who try to answer questions like these do so by applying a "discount rate" to their calculations. The higher the positive discount rate, the less valuable future consumption is considered to be vis-à-vis present consumption. Most energy decisions are made using fairly high positive discount rates. For example, a barrel of oil today is valued much more highly than a barrel of oil scheduled for delivery a year from now. A barrel of oil one hundred years from now has essentially no present value. Little oil may be left in one hundred years, but the economists assume that something else—such as synthetic fuels derived from coal, or chemicals made from trees—will have replaced it. In fact, of course, nothing may have satisfactorily replaced it, and in 2076 our great-grandchildren might be willing to pay a great deal for a barrel of oil. However, since no one is willing to buy that barrel of oil today and to set it aside for them, it will instead be bought for $13 and consumed immediately. This price, although five times as high as the prevailing price a few years ago, is still low enough to ensure that global oil production will peak and begin its decline during our lifetime.

Of course, we are not in a free market situation. The oil cartel, OPEC, has already decided not to produce oil as rapidly as is physically possible. OPEC prefers to act as a rational, farsighted monopoly—translating long-term scarcities into short-term scarcities. At some point, some of the oil-rich nations will find themselves with more income than they can reasonably spend or invest. Indeed, some prominent Saudi Arabians feel their country has already passed this point.

In addition to economic discounting, energy resource decisions are influenced by what might be termed "political discounting." Many elected politicians consider the next election to be the most important of all horizons; anything that produces ill effects beyond the next election matters little. Thus, all tax cuts precede elections, and consequent inflation follows them. Votes are won by ensuring the greatest possible current prosperity at the lowest possible prices, and political decisions that impede consumption are exceedingly rare, while those that encourage rapid exploitation are the rule. The jingle of the cash register can drown out the voices of the unborn.

While the world as a whole faces no current shortage of fossil fuels, those areas in which energy demands have already outstripped domestic supplies should immediately begin a transition toward use of renewable sources. With only slightly less urgency, the remainder of the world should follow suit. Unless we undergo a revolutionary change of direction, 80 percent of all the oil and gas on earth will be consumed by the current generation. The cry *"Après moi le déluge"* sounds as insane coming from a single generation as from a single monarch.

3. Nuclear Power: The Fifth Horseman

IN THE 1950s and early 1960s, the U.S. Air Force invested over $1 billion attempting to build a nuclear-powered airplane. Some critics pointed out that it would be too heavy and cumbersome to be militarily useful, others that radioactive debris would be scattered over the countryside if the plane crashed. Still the Air Force pushed relentlessly on until 1962, when President Kennedy finally ordered the project scrapped.

For two decades, commercial nuclear power has grown steadily, spreading to more than twenty countries. It has acquired strong advocates in corporate boardrooms, labor union headquarters, and governmental energy bureaucracies. Nonetheless, a potent worldwide political constituency has come to view commercial nuclear power as President Kennedy viewed the nuclear airplane—an idea that just isn't going to fly.

In the mid-1950s, the United States, the Soviet Union, Britain, and France all began operating nuclear reactors to generate electricity. The Federal Republic of Germany began reactor operations in 1960, Canada and Italy joined the club in 1962, and Japan and Sweden followed in 1963. Also in this period, the People's Republic of China began limited weapons-related reactor operations, exploding its first nuclear bomb in 1964.

By 1970, the list of nations with commercial nuclear facilities had lengthened to include Switzerland, the German Democratic Republic, the Netherlands, Spain, Belgium, and India. Since then, Pakistan, Taiwan, Czechoslovakia, Argentina, and Bulgaria have joined the ranks, bringing the total to twenty-one. In 1976, nuclear power accounted for 21 percent of all electricity generated in Belgium, 18 percent in both Sweden and Switzerland, 13 percent in Great Britain, and 9.4 percent in the United States.

By 1977, the world's 204 commercial reactors had a combined capacity of 94,841 megawatts of electricity—up more than tenfold in ten years. Planned additions would quickly multiply that capacity almost eightfold to 569,544 megawatts, derived from 682 reactors. By the end of the century, fifty or more countries could have a combined generating capacity of more than 2 million megawatts.[1] However, such development is beginning to look exceedingly unlikely.

In much of the industrialized world, the future of "the peaceful atom" has grown cloudy. In the spring of 1973, the Swedish Parliament called a halt to nuclear power development while the government initiated a public education program. By the time of the final governmental decision on May 29, 1975, a majority of Swedes opposed the construction of more reactors. A parliamentary coalition voted to limit future nuclear construction to two reactors beyond those already planned at the time of the moratorium.[2] In September of 1976, a strongly anti-nuclear new prime minister, Thorbjorn Fälldin, was elected.

The number of reactor orders annually placed in the United States reached a peak of 36 in 1973, declined to 27 in 1974, and plummeted to 4 in 1975. As of mid-1976, no new reactors have been ordered. Indeed, cancellations and deferrals outpaced new reactor orders in the United States by more than 25 to 1 in 1975. Even as numerous states debate nuclear moratoria and other restraints, a *de facto* national moratorium appears to be in effect.

Nuclear development has hit shoals all around the world. In Japan, it has been snagged by a series of lawsuits and by widespread protest rallies. Japan's first nuclear-driven ship, the *Mutsu*, developed a widely publicized radiation leak during a trial run in September of 1974. To the south, an Australian coalition of environmental groups and trade unions has brought nuclear development to a standstill. Australia has no plans to build domestic reactors, and the public is debating whether the country should even export uranium. Widespread nuclear opposition has also surfaced in England, France, Germany, Austria, Denmark, and New Zealand, and evidence suggests that quiet opposition exists inside the Soviet Union.

The Canadian government continues to laud the virtues of its CANDU (Canadian Deuterium Uranium) reactor, but public opposition has mounted rapidly in recent years. Much opposition arose in response to India's decision to construct nuclear explosives out of plutonium produced in a reactor supplied by Canada.

In the early 1970s, as nuclear construction faltered in much of the developed world, nuclear vendors turned to less industrialized countries. Corporations seeking to recoup enormous research investments entered into fierce competition for Third World reactor orders. Yet, for most poor countries, a capital-intensive, highly centralized, and technically complicated source of electricity is a tragically inappropriate investment.

A generally accepted guideline is that no single power plant should represent more than 15 percent of the capacity of a power grid. Otherwise, the shutdown of a single power plant can impair the entire system. By this rule of thumb, only those countries having at least 4,000 megawatts of installed capacity on a single transmission network should even consider a single small (600-megawatt) reactor. Argentina, Brazil, Egypt, India, Korea, Mexico, and Venezuela are the only developing countries that could currently support even one such nuclear plant. Nuclear vendors are hungry for new markets, however, and are therefore willing to offer much more liberal credit arrangements than would generally be available for alternative technologies. The U.S. Export-Import Bank, for example, has made loans of about $3 billion in support of American nuclear sales in eleven countries. The largest credit ever approved by the Eximbank was in support of the recent sale of a Westinghouse reactor to the Philippines.[3]

International nuclear sales are generally made on the pretext of fostering energy independence. But far from freeing poor countries from OPEC's influence, nuclear power will make poor countries even more dependent upon rich ones for fuel and technology, since the global distribution of high-grade uranium ore is even less equitable than the distribution of oil. Eighty-five percent of non-Communist uranium reserves are concentrated in just four countries: the United States, Canada, South Africa, and Australia. Access to enrichment and reprocessing technologies appears certain to be increasingly restricted.[4] And nuclear power is incomparably more complex and less labor-intensive than other energy sources. As the Third World comes to appreciate fully the social and economic consequences of nuclear development, this growth market is likely to become limited to only those nations who seek commercial nuclear power as a step toward nuclear armaments.

In recent years, many nuclear problems have been widely debated. Nuclear opposition originally arose during a dispute over the carcinogenic properties of ionizing radiation. With the passage of time, nuclear opponents expanded their attacks to encompass problems of waste disposal, economics, fuel availability, and the safety of breeder reactors. The literature on these issues fills volumes and grows daily. Several comprehensive reviews exist, and this discussion will therefore be limited to a brief description of the crux of each argument.

Three new issues, however, warrant more attention. Although they have not figured prominently in most national nuclear debates, all are of paramount importance internationally, and none appears to have a technical solution. First, the proliferation of commercial nuclear power will almost inevitably lead to the widespread possession of nuclear weapons. Second, it will heighten humankind's vulnerability to terrorism. And, third, it will foster the evolution of highly centralized technocratic and authoritarian societies.

Radiation

The environmental threats posed by the nuclear power cycle cannot be fully measured without an understanding of the effects of radiation on life at the molecular level—an understanding that is at present far from complete.[5] The radiation associated with nuclear power is emitted through the spontaneous decay of reactor-produced radioactive materials. In addition to its 100 tons of uranium oxide fuel, one large modern reactor contains about two tons of various radioactive isotopes—one thousand times as much long-lived radioactive material as the Hiroshima bomb produced.

As sub-atomic particles of radiation (X rays, gamma rays, alpha particles, beta particles, and neutrons) shoot out from decaying atoms, they collide with other matter, generally with electrons. In such collisions, so-called "ionizing radiation" frequently jars the electron free from the atom of which it is a part; this electron loss transforms the atom into a positively charged ion.

Nuclear industry workers are exposed to more radiation than is the general public. The need to make repairs on radioactive equipment poses a particularly intractable risk. Any single worker can tolerate only brief exposure; as many as six men have reportedly been required to remove one nut from one bolt. Consolidated Edison, a New York utility, required a few minutes of work from each of 1,500 skilled workers to weld and insulate six hot-water pipes at its Indian Point Number One plant. When an accident partially destroyed the core of Canada's Chalk River facility in 1952, one of the imported technicians—each of whom worked ninety seconds at the irradiated Chalk River reactor—was a young American navy officer and nuclear engineer named Jimmy Carter.

Should a nuclear accident occur, however, the public as well as the workers could be imperiled by radionuclides. Even routine emissions from a normally functioning fuel cycle may pose dangers. Lacking an understanding of the molecular effects of radiation, we don't even know whether very low exposures cause damage or whether there is a threshold below which exposure to radiation is harmless. Nuclear advocates say that no danger has been proven; nuclear critics respond that safety has not been proven. Both are correct.[6]

Radioactive Waste

No country has yet devised an adequate solution to the problems posed by high-level radioactive waste. Such waste is of two basic types: fission products and actinides. Fission products, which include strontium 90, cesium 137, and krypton 85, are produced when atoms of uranium or plutonium are split in reactors. The principal fission products have half-lives of thirty years or less, so 700 years from the time they are produced only a negligible one ten-millionth remains. Actinides, such as actinium, neptunium, americium, and einsteinium, are formed when atoms of uranium or thorium absorb neutrons from the splitting of fissile fuels. All actinides are highly toxic and have exceedingly long half-lives. The most common actinide, plutonium 239, has a half-life of 24,700 years. The actinides are more toxic but much less radioactive (for the first 500 years or so) than the fission products.

The principal nuclear waste accident to date occurred in 1958 at the Soviet repository in the Ural Mountains near Blagoveshchensk. An unexplained explosion blew radioactive materials sky-high, and strong winds distributed them over hundreds of miles. Soviet biochemist Zhores Medvedev writes, "Tens of thousands of people were affected, hundreds dying . . . ," though the Soviet government has never officially admitted the incident.[7]

Most waste strategies are based upon the assumption that all types of high-level wastes will be disposed of together. For the time being, wastes are kept in surface repositories from whence they occasionally leak, to the consternation of people living in adjacent areas. Radioactive wastes from U.S. military operations have proven particularly troublesome. More than 400,000 gallons have leaked from the waste repository at Hanford, Washington; smaller leaks have occurred at the Savannah River facility in Georgia.

All long-term disposal strategies reflect the assumption that high-level wastes will eventually be stored in solid rather than liquid form. Mixed with twice its volume of inert material in a glasslike solid, the high-level waste from a 1,000-megawatt reactor fills about 100 cubic feet a year. The United States plans to store such waste in steel canisters, each of which measures 3 meters long and 0.3 meters in diameter. If current growth projections prove true, the American nuclear industry could produce 80,000 such canisters over the next twenty-five years.

Orbiting satellites, arctic ice caps, and deep salt mines have been suggested as permanent repositories for nuclear waste. The United States government was forced to abandon its plan to create a dump for high-level nuclear wastes near Lyons, Kansas, after the local salt mine proved to have copious leaks. Salt-bed storage is currently being investigated by West Germany and Canada, while Sweden is experimenting with disposal in granite and Italy favors disposal in clay.

Even low-level nuclear waste is proving troublesome. The volume of low-level waste scheduled for production in the United States alone by the year 2000 will, according to the U.S. Environmental Protection Agency, amount to about one billion cubic feet—enough to cover a four-lane coast-to-coast highway one foot deep.[8]

Burial grounds for low-level waste have been selected without first making hydrological and geological studies. Moreover, according to a disturbing study by the U.S. General Accounting Office, "there is little or no information available on the chemical or physical nature of the wastes." In early 1976, the U.S. Environmental Protection Agency found plutonium percolating through the soil at the burial grounds for low-level waste at Maxey Flats, Kentucky.[9]

Much low-level radioactive waste is currently cast into the ocean. Before 1967, this dumping went unsupervised. Between the mid-1940s and the mid-1950s, the United States occasionally dumped radioactive rubbish into both the Atlantic and the Pacific oceans, while Britain has used the Atlantic as its dumping ground. Controls have been gradually strengthened since the mid-1960s, but the problem persists. In 1975, the Nuclear Energy Agency supervised the dumping of 4,500 tons of low-level European nuclear waste into the Atlantic, 1,300 kilometers due west of France. These drum-packaged wastes joined 34,740 tons of nuclear waste previously dumped at this location.[10]

Nuclear Economics

Global nuclear development was initially spurred by the belief that fission would provide a cheap, clean, safe source of power for rich and poor alike. However, the dream of "electricity too cheap to meter" has foundered.

Nuclear power is not cheap. Donald Cook, chairman of American Electric Power—the largest privately owned utility system in the United States—believes that "an erroneous conception of the economics of nuclear power" sent U.S. utilities "down the wrong road. The economics that were projected but never materialized—and never will materialize—looked so good that the companies couldn't resist it."

The costs of nuclear power are mostly at the front end—in research and development and capital construction. Consequently, such power facilities will necessarily be at a severe disadvantage in a time of general capital scarcity. And while all capital costs have been increasing dramatically in recent years, the cost increases of nuclear construction have outpaced the rises in the construction costs of other power facilities. The per kilowatt price of U.S. nuclear facilities rose two-and-one-half times as much between 1969 and 1975 as did that for coal-fired power plants.[11]

The true cost of nuclear power has been confused by the quasi-public nature of much nuclear research and development. The costs of decommissioning radioactive facilities, the costs of regulation (including effective safeguards), and the cost of safe disposal of wastes are all generally ignored. Moreover, the typical reactor produces power at just over one-half of its designed capacity, owing to shutdowns and slowdowns for safety reasons. A study of nuclear costs by physicist Amory Lovins revealed that nuclear power requires a total investment of $3,000 per kilowatt of net, usable delivered electric power. In other words, lighting a single 100-watt bulb by nuclear power requires a $300 investment.[12]

Projected nuclear growth in the United States through the year 2000 could require more than one-fourth of the nation's entire net capital investment. In some developing countries, the cost of a single reactor may exceed the amount of the nation's total annual available capital. Such investments represent grievously injudicious use of scarce capital.

Uranium Availability

Uranium is not a plentiful substitute for scarce oil and gas. Total non-Communist uranium resources available at $60 per kilogram have been estimated in a 1975 study by the OECD Nuclear Energy Agency and the International Atomic Energy Agency (IAEA) at about 3.5 million tons—about half of which was reasonably assured. Three countries control 80 percent of current non-Communist production: the United States, with 9,000 tons per year; Canada, with 4,700 tons; and South Africa, with 2,600 tons. Eighteen other countries have discovered small uranium deposits, but the total from these countries represents only 15 percent of the non-Communist resource base. (Public information is not available on the uranium resources of the Soviet bloc or of the People's Republic of China.)[13]

The 236 reactors currently operating or planned for construction in the United States will consume at least 1 million tons of uranium oxide over their lifetime. The 800 U.S. reactors sometimes projected to be in operation by the year 2000 will cumulatively demand over 2 million tons through that year, and will demand 4 million tons altogether during their operating span. These fuel demands—projected by the U.S. Energy Research and Development Administration, and challenged as far too low by others—outstrip the economically recoverable reserves of all known non-Communist uranium suppliers.

What holds true for the United States is, in this instance, even more emphatically true for the world. While cumulative demand for uranium oxide in the United States could total 2 million tons by the year 2000, cumulative non–U.S. demand is expected to exceed that amount. Proposed non–U.S. reactors will themselves have a lifetime demand far in excess of the world's known deposits of economical uranium. Low-cost ores over and beyond those now postulated may well be unearthed; on the other hand, most of the estimated resource base is hypothetical, and actual deposits could easily fall short of the estimates.

Without breeder reactors, known uranium reserves obtainable at reasonable prices will not long support nuclear development. Of course, as prices rise, the amount of uranium recoverable will also rise. But exploiting low-grade ore incurs heavy non-economic costs. In the United States, uranium is now mined from western sandstone, in which it comprises 1,000 parts per million. In the lower-grade Chattanooga shale, uranium constitutes only 60 to 80 parts per million—less uranium than the tailings currently being discarded from uranium milling operations. Of that minuscule amount of uranium, less than 1 percent is fissionable U 235; the rest of the uranium cannot be split to release energy.

The energy cost of extracting so little fissile fuel from so much ore may topple the nuclear industry. Although one preliminary study suggests that a net energy gain is still possible, such a gain may not be worth the effort and may not represent a judicious investment of manpower and capital. Ton for ton, Chattanooga shale contains less energy than does bituminous coal, and the environmental costs of uranium extraction from such ore will be high.

Reactor Safety[14]

A 1,000-megawatt reactor, after sustained operations, has about 15 billion curies of radioactive material in its core. The heat of decay from this material constitutes about 7 percent of the reactor's thermal output (the other 93 percent coming from the fission reaction).[15] While the fission process can be regulated, radioactive decay cannot. The decaying core can only be cooled. Uncooled, the core would grow so hot that it could melt through its containment vessel, and would then continue to melt its way down into the earth. This "loss of coolant accident" (LOCA) has been the focus of most of the reactor safety controversy.

There is no question but that such accidents can occur. The questions, rather, are how dangerous a meltdown would be and how frequently a meltdown would be likely to occur.

A once secret 1957 report prepared by the Brookhaven National Laboratory for the U.S. Atomic Energy Commission concluded that the worst possible reactor meltdown could kill 3,400 people, injure 43,000, and cause $7 billion damage. By 1964, larger reactors were on the market and an updated Brookhaven report upped the estimated toll, claiming that 27,000 people could die, that $17 billion worth of damage could be done, and that an area the size of Pennsylvania could be contaminated. A study conducted by the Engineering Research Institute of the University of Michigan for the owners of the Enrico Fermi reactor outside Detroit found that the worst accident likely to occur with this relatively small breeder reactor could cost 133,000 lives.

None of these studies dealt with the odds of such an accident occurring. In 1972, the United States AEC sponsored yet another reactor safety study.[16] Known by the name of its principal author, the Rasmussen study traced the sequences of events that could—as the analysts saw it—lead to a LOCA, and assigned a probability to each event and then to the sequences. The Rasmussen report claims that a core meltdown will occur about once every 17,000 reactor years for pressurized water reactors, and about once every 33,000 years for boiling water reactors. These calculations reflect the presumption that neither God nor terrorists will intervene with unscheduled events and the belief that Rasmussen's thousands of assumptions about reactor components are all correct. For example, the report maintains that the emergency core cooling system (ECCS) will work successfully unless some pump, valve, or other component fails. However, many experts doubt if the ECCS can prevent a meltdown *even when working perfectly*, and the system has never been tested.[17]

Doubtless, the most publicized result of the Rasmussen study was a chart comparing the relative odds of a person dying from a nuclear accident, being struck by lightning, being struck by a meteor, and so on. Nuclear power, unsurprisingly, was found to be wondrously safe. The catch, however, is that these charts consider *immediate* deaths only. Professor Frank von Hippel of Princeton University points out that an accident that causes only 10 early fatalities by Rasmussen's calculations would subsequently cause 7,000 cancer deaths, 4,000 genetic defects, and 60,000 thyroid tumor cases. It would also contaminate 3,000 square miles of land.

Most of the immediate danger to human life posed by a serious reactor accident arises from the cloud of radioactive material that would be released if the reactor containment vessel were breached. The number of people exposed would depend upon the population density in the surrounding area, upon climatic conditions, and upon the effectiveness of evacuation procedures. Sixteen million people live within a forty-mile radius of the three reactors at Indian Point, New York. In February, 1976, Robert Pollard, the safety official directing regulatory activities at Indian Point, resigned and announced on national television that Indian Point Number Two was "almost an accident waiting to happen."

The likelihood of a successful rapid evacuation of a congested area containing several million people is equal to that of an apple falling upward, and this is frankly admitted by the state officials. "What's my plan to evacuate Chicago?" asks the nuclear chief of the Illinois Office of Civil Defense. "I don't have one. There's no way you can evacuate Chicago." In few reactor accidents has the public even been informed that a potential danger existed until after the critical period had passed. The head of civil defense in the Browns Ferry area didn't hear about a $100 million fire that incapacitated two 1,100-megawatt reactors until two days after the fire was put out.[18]

In November of 1973, a Swedish radio program describing a fictional reactor accident in southern Sweden was broadcast. The resulting public panic recalled the shock created by Orson Welles' *The War of the Worlds* some four decades earlier. The phone system broke down under the stress of calls, within ten minutes an enormous traffic jam had tied up the countryside, and frantic citizens were reluctant to believe official assurances that no accident had taken place.

The nuclear safety debate has been a source of great confusion to the layman. One team of experts is lined up against an equally expert opposing team, each armed with computer printouts and technical jargon. Each tries to "prove" its case. But most nuclear issues are not amenable to proof; they are matters of judgment. It is impossible to eliminate all risk, and determining the level of acceptable risk is an ethical rather than a technical exercise. Consequently, the final decisions are not scientific, but are, rather, social, political, and philosophical.[19]

Breeder Reactors

Rhapsodie Fortissimo, Phoenix, and SNEAK are some of the names given to prototypes of an exotic new technology that would produce more fuel than it consumes. Breeder reactors perform a certain alchemy, transforming atoms with no potential as fuels into entirely different elements whose energy can be exploited. The leading breeder candidate is the liquid metal fast breeder reactor (LMFBR), designed to transform uranium 238 (the non–chain-reacting isotope that constitutes more than 99 percent of all uranium) into plutonium 239, a reactor fuel. Other proposed breeders would convert thorium into fissionable uranium 233.[20]

The "doubling time"—the amount of time needed for a breeder reactor to accumulate twice as much fissionable fuel as its initial inventory contained—is a critically important aspect of breeder development. The more rapid the doubling time, the larger the amount of useless U 238 the breeder will convert into valuable plutonium 239 during a given operating period. Because the breeder converts otherwise valueless material into fuel, it in effect increases the size of the uranium resource base: more energy is obtained per unit of fuel mined, and lower grades of fuel can be economically mined. If nuclear fission is viewed simply as a stopgap or supplementary power source, the meager known resource base of fissile fuels may be adequate, and the breeder may be justifiably characterized as an expensive extravagance. If, on the other hand, nuclear fission were to become a major long-term energy option, breeder reactors—with all their attendant problems—would be indispensable.

Fast neutrons cause a vast atomic stir inside a LMFBR. This neutron bombardment creates voids in the crystalline structure of metallic fuel rods, swelling both the metal cladding and the fuel itself as a consequence. If fuel pins bow and touch as a result of this swelling, temperatures increase greatly at the contact points. Under some circumstances, this heat could spread to other parts of the core and initiate melting. The current breeder safety debate centers on whether or not the fuel could become arranged in an explosive configuration during a core melt (a condition known as "recriticality") and blow the reactor apart (or, in technical jargon, cause a "rapid disassembly"). Just how much energy such an explosion would release is not known.[21]

The easiest "solution" to the swelling problem is to design more space (filled with sodium) between the fuel pins so that, even if they bend, they won't touch. However, the sodium flowing between the pins slows down the neutrons and reduces the breeding rate. The contribution of the breeder to fuel supplies will be marginal unless the breeding time is brought down substantially from the present forty-to-sixty-year range. Thus, safety and speed are at loggerheads, for a cut in the breeding time will require a closer fitting of fuel pins unless there is a breakthrough in fuel technology.

In October of 1966, instruments on the Enrico Fermi reactor in Lagoona Beach, Michigan, began to behave erratically. An LMFBR, Fermi was the world's first commercial breeder reactor. Suddenly, the reactor's radiation warning device registered an emergency. It was impossible to tell what was occurring in the reactor core, but the instrument readings supported the hypothesis that at least one fuel subassembly had melted. Safety was of special concern at Fermi because 4 million people resided within thirty miles of the reactor.

The Fermi reactor was successfully shut down. During the next several days, experts were flown in from all over the world to speculate upon what might be happening in the reactor's core. The greatest fear was that a damaged subassembly might collapse into other parts of the core, causing a secondary nuclear accident of catastrophic dimension. Slowly, the delicate operations were begun. More than a year and a half of careful work was required before the cause of the accident could be discovered: a triangular piece of metal installed as a safety measure had worked loose, clogging the flow of coolant and causing four fuel subassemblies to melt. Tragedy was only narrowly averted.

Perhaps the greatest fear that breeder reactors inspire is that nothing will go wrong, that the plants might be commercialized in a timely manner and in an economical form, and that they might operate without mishap. In this case, the world could come rapidly to depend upon plutonium as a principal fuel. Some consequences of such an unholy addiction will be explored in the next three sections.

Weapons Proliferation

In August, 1939, Albert Einstein wrote a letter to President Franklin D. Roosevelt of the United States. "Some recent work by E. Fermi and

L. Szilard which has been communicated to me in manuscript form leads me to expect that the element Uranium may be turned into a new and important source of energy in the immediate future."

The letter led to the Manhattan Project—a multinational undertaking that gave birth to the first atom bomb. Some idealistic supporters of the project dared to believe that their efforts would lead to world peace. With the threat of nuclear weaponry looming grotesquely in the background, war would become unthinkable.

Since the explosion of the first nuclear device, the world has experienced scores of regional wars, and has twice set foot on the brink of nuclear holocaust. During this period, the international nuclear arsenal grew to absurd proportions, desecrating the hope that our future will be less war-torn than our past.

Today all five permanent members of the UN Security Council have exploded nuclear bombs. So has India. Approximately fifteen more countries are in what could be termed "near nuclear" status; they could, no doubt, quickly produce nuclear weapons if they chose to do so.[22]

Virtually all nations agree that the widespread dissemination of nuclear armaments would gravely jeopardize not only global stability but perhaps even the survival of the human species. In the event of an accidental or intentional nuclear war, the incredible impact of the initial conflagration (the world's nuclear arsenals today contain the equivalent of 20 billion tons of TNT) would be followed by long-term radiation damage, ozone depletion, and, possibly, major climatic shifts. Our ignorance of the effects of such a massive assault on the global environment is nearly total.[23]

After the Cuban missile crisis of 1962, the United States and the USSR became more acutely aware of the fragility of the nuclear age. The following year, the Limited Nuclear Test Ban Treaty was signed. In 1967, the Treaty of Tlatelolco prohibited the development of nuclear weapons in Latin America. And on March 5, 1970, the Treaty on the Non-Proliferation of Nuclear Weapons (NPT) went into effect.

Written by the United States and the Soviet Union, the NPT treaty makes a good deal of sense from a superpower perspective. Both countries retain their vast arsenals, and each continues to manufacture about three hydrogen bombs a day. Non-weapons states, however, are prohibited by the treaty from developing or acquiring nuclear weapons. Non-weapons states are subjected to IAEA inspections; the nuclear powers are not. The superpowers' sole obligation is to make good faith efforts toward nuclear disarmament. Virtually no non-nuclear power believes that such efforts have actually been made.[24]

"If I had known in 1968 how little the nuclear powers would do over the next six years [to control the arms race]," remarked one highly placed senior diplomat of a non-nuclear country, "I would have advised my government not to sign the treaty." Countries that have not signed the treaty include India and Pakistan, Argentina and Brazil, Egypt and Israel, China, South Africa, and France.

The regrettable fact is that the NPT offers nothing, or less than nothing, to its non-weapons participants. None of the nuclear exporting nations is willing to limit its nuclear exports to states agreeing to place all their nuclear activities under IAEA safeguards; none wishes to lose a potential sale. Thus, parties to the NPT voluntarily relinquish a degree of sovereignty, while non-parties have nuclear vendors beating down their doors with offers of nuclear hardware.

The general disillusionment with NPT may be gauged by the record of the long-awaited Five-Year Review Conference held in Geneva in May, 1975. The prelude to the conference deserves note. India had detonated her first nuclear device on May 18, 1974. In June of that year, the American president offered 600-megawatt reactors to Egypt and Israel—two fiercely antagonistic non-NPT states. And the 1974 Vladivostok agreement between the United States and the Soviet Union—far from upholding the superpowers' NPT obligations to bring the arms race to a timely conclusion—was widely perceived as a slightly modified set of ground rules for the continuation of that race.[25]

Some of the flavor of the Geneva conference may be captured by tracing the fate of an exceedingly modest proposed protocol under which the nuclear powers would have agreed not to use nuclear weapons against countries not having nuclear weapons, to assist non-nuclear countries that were threatened or attacked with nuclear weapons, and to encourage negotiations to establish nuclear weapon–free zones. The nuclear powers refused this protocol out of hand—a traditional posture for the United States, but a new one for the Soviet Union. Thus, non-weapons countries that agreed to become parties to the Non-Proliferation Treaty were unable to obtain assurances that the nuclear powers would not launch nuclear strikes against them! At about this time, James Schlesinger, the U.S. Secretary of Defense during the Nixon administration, publicly reaffirmed his nation's willingness to use nuclear weapons in response to a conventional attack.

The nuclear weapons states at the conference dismissed all proposals made by developing nations, calling such proposals "political" in nature, and urged instead that the conference limit itself to the technical problems of NPT implementation. By this, they meant the strengthening of safeguards on nuclear material. But the nuclear powers provided no concrete proposals as to how security might be tightened. They supported the concept of international nuclear power centers, but offered only vague ideas about how these might be handled. Regional centers able to serve Argentina and Brazil, India and Pakistan, Israel and the Arab states struck many observers as problematical.

The conference, viewed from any perspective, was a failure. Shortly after the meeting adjourned, West Germany announced its $4 billion sale of a complete nuclear fuel cycle to Brazil, a non-party to the NPT. Brazil had already proclaimed its intent to develop nuclear explosives for "peaceful purposes" only, but Fred Ikle, head of the U.S. Arms Control and Disarmament Agency, has noted that a very sophisticated warhead could be tested in a "peaceful" explosion designed to build a dam.[26]

Adherence to the NPT holds no advantage for any country other than a superpower, and development of nuclear explosives arguably does. China, virtually ignored by other governments until it exploded its bomb in October of 1964, has since obtained a seat on the UN Security Council and has become a respected force in the community of nations. The Indian bomb, far from eliciting international opprobrium, evoked only a spate of political cartoons and short-lived censure from two or three countries. In India, the explosion greatly strengthened the domestic stature of the ruling Congress Party and of its leader, Indira Gandhi. U.S. Secretary of State Henry Kissinger, visiting India five months after the blast, asked only that India act responsibly on the export of nuclear technology. Small wonder that in April of 1975, while introducing a bill calling upon his country to construct an atom bomb, one Argentinian legislator stated that "recent events have demonstrated that nations gain increasing recognition in the international arena in accordance with their power."

The existence of nuclear weapons in some lands leads almost inexorably to their development in others. The Chinese bomb arguably spawned the Indian device, and the Indian explosion seems likely to beget a Pakistani bomb. Pakistani Prime Minister Zulfikar Ali Bhutto growled that he will "never surrender to any nuclear blackmail by India. The people of Pakistan are ready to offer any sacrifice, and even eat grass, to ensure nuclear parity with India." Even among the Japanese—the only people ever to have suffered a nuclear attack—a broad consensus holds that the advent of a Korean bomb would turn Japanese antinuclear public opinion around overnight. Israel is widely believed to have between ten and twenty small nuclear weapons. South Africa is also thought by some to possess a modest nuclear arsenal. The ruling military governments in many lands are no doubt aware of the strategic significance of nuclear weapons.

There is almost certainly a threshold number of nuclear nations, the existence of which would serve to convince holdout countries that continued abstinence is purposeless. At that point, wherever it is, the NPT dam will break and the world will go nuclear. "I'm glad I'm not a young man, and I'm sorry for my grandchildren," says David Lilienthal, the first chairman of the U.S. Atomic Energy Commission. Such concerns can only deepen: the reactors that U.S. manufacturers alone plan to sell internationally over the next decade will produce enough plutonium *each year* to make 3,000 small bombs.

With so many near-nuclear states not parties to the NPT, with the future of that treaty clouded by uncertainties, and with the nuclear exporting countries engaged in fierce competition for international markets, the future worth of the IAEA safeguards program is highly questionable. However, if only because nuclear proponents generally express great confidence in IAEA policing activities, the safeguards program requires a brief examination.

Conceded by even its strongest admirers to be a shoestring operation, the IAEA safeguards program conducts inspections in 92 NPT countries and in non-treaty states that have agreed to such inspections. (All nuclear vendors except France now demand such inspections as a condition of sale.) To accomplish this trying task, the IAEA employs 70 technicians and has a budget of about $5 million. The organization's primary regulatory activity is auditing records. The occasional on-site examinations it sponsors are ordinarily announced well in advance.

Besides its exceedingly modest scale and budget, four other major problems hamstring the IAEA. First, a nation violating its commitments would have to be remarkably inept to be caught in an auditing error. When volumes of fissile materials are large, even a small margin of uncertainty can lead to significant losses; and bomb-sized gaps are simply not covered by existing safeguards. One percent of a pound of plutonium won't make a bomb, but one percent of a ton will. When material is converted to and from gaseous, liquid, and solid states—as the fuel cycle requires—losses and inaccuracies are inevitable. The United States probably has the finest nuclear safeguards program in the world, yet cumulative U.S. losses of fissile material could fill an enormous arsenal. The most significant losses occurred in the early years of the nuclear program, but, as recently as December of 1975, a fuel fabrication plant in Erwin, Tennessee, reported an auditing discrepancy involving 20–40 kilograms (44–88 pounds) of fully enriched uranium.

The second problem with the international safeguards program is that coups, revolutions, and other government upsets will often invalidate all agreements made by previous leaders. The United States flew a secret team of experts into South Vietnam to de-fuel and then demolish that country's only reactor shortly before the fall of the Thieu regime.

A third weakness of the NPT safeguards program is that the IAEA has no authority to take any action against violations other than to announce them. Indeed, most countries consider occasional inspections to impinge upon their sovereignty; few, if any, would grant an international police team the authority to confiscate bomb-grade material.

Finally, selling hardware necessarily means selling knowledge. Sales of nuclear hardware are subject to safeguards, but duplicate facilities built by the recipient countries will not be. Brazil, for example, is less apt to build a bomb by sneaking material out of the German-built facilities than it is to openly build similar facilities of its own for the avowed purpose of developing peaceful nuclear explosives. Brazil's rival, Argentina, has ordered a large CANDU reactor from Canada. The Canadian government required a pledge that CANDU-produced plutonium would not be used for weapons. "It's really a little silly," states a spokesman for the Argentine Embassy in Ottawa. "We'll sign the agreement all right. But how do they expect to enforce it? Besides, we wouldn't dream of building a nuclear bomb—unless Brazil does."

Six countries have now exploded nuclear devices. At least fifteen other countries have the fissile materials and the technical competence to manufacture bombs. Widespread weapons proliferation is sure to follow the rapid growth of commercial nuclear power facilities.

Nuclear Terrorism

Three materials with weapons potential play prominent roles in nuclear power fuel cycles. Plutonium 239, made inside all existing commercial reactors, is highly toxic, carcinogenic, mutagenic, and explosive. Uranium 235 is the fuel of most existing commercial reactors, and uranium 233 is produced in reactors containing thorium. Spheres of Pu 239, U 235, and U 233, encased in a beryllium neutron reflector, have critical masses of 4 kilograms (under 9 pounds), 11 kilograms, and 4.5 kilograms, respectively.[27] Sophisticated implosion techniques can lower the critical mass requirements considerably; for plutonium used in implosion bombs, the official "trigger quantity" is about 2 kilograms. A skilled bombmaker would require slightly less than these official figures suggest. An amateur bombmaker could make a less sophisticated weapon employing correspondingly larger amounts of fissile material. A recent report by the "watchdog" agency of the U.S. Congress, the General Accounting Office, found that "even minimal and basic security precautions had not been taken" to protect plutonium. The report cited, among other examples, "an unlocked and unalarmed building containing plutonium scrap . . . within 15 feet of an unalarmed fence."[28]

Until 1970, the United States government purchased all the plutonium produced in U.S. reactors. In 1970, the government got out of the business, and private companies began stockpiling the material. If reliance on nuclear power grows at the rate commonly projected, far more plutonium will be produced in commercial reactors in the next couple of decades than is now contained in all the nuclear bombs in the world. Theodore Taylor, a nuclear safeguards expert, estimates that by the year 2000 enough fissile material will be in circulation to manufacture 250,000 bombs. If U.S. Atomic Energy Commission growth projections for nuclear power through 2020 were to be met, Arthur Tamplin and Thomas Cochran have calculated, the cumulative flow of plutonium in the United States alone would amount to 200 million kilograms (440 million pounds).

Once assembled, nuclear weapons could be rather convenient to use. The dimensions of the Davy Crockett, a small fission bomb in the U.S. arsenal, are 2 feet by 1 foot (0.6 meters by 0.3). The smallest U.S. bomb is under 6 inches (0.15 meters) in diameter. Such bomb miniaturization is well beyond the technical skill of any terrorist group, but no wizardry is required to build an atom bomb that would fit comfortably in the trunk of an automobile. Left in a car just outside the exclusion zone around the U.S. Capitol during the State of the Union address, such a device could eliminate the Congress, the Supreme Court, and the entire line of succession to the presidency.

With careful planning and tight discipline, armed groups could interrupt the fuel cycle at several vulnerable points and escape with fissile material. The high price likely to be charged for black market plutonium also makes it attractive to organized crime: sophisticated yet ruthless, modern criminals have close links with transport industries in many parts of the world. Perhaps most frightening is the inside thief—the terrorist sympathizer or the person with gambling debts or the victim of blackmail. A high official of the U.S. Atomic Energy Commission had, it was discovered in 1973, borrowed almost a quarter of a million dollars and spent much of it on racing wagers.

Quiet diversion of bomb-grade material may have taken place already. Plutonium has often been found where it should not have been, and, worse, not been found where it should have been. Determining whether or not weapons-grade material has already fallen into the wrong hands is impossible. Charles Thornton, former director of Nuclear Materials Safeguards for the U.S. Atomic Energy Commission, claims that "the aggregate MUF [materials unaccounted for] from the three U.S. diffusion plants alone is expressible in tons. No one knows where it is. None of it may have been stolen, but the balances don't close. You could divert from any plant in the world, in substantial amounts, and never be detected. . . . The statistical thief learns the sensitivity of the system and operates within it and is never detected."

It was long and incorrectly believed in the United States, as it is still believed elsewhere, that building a bomb from stolen materials would require "a small Manhattan project." But Theodore Taylor, formerly the leading American atom bomb designer, has described at length where the detailed instructions for building atomic bombs can be found in unclassified literature and how the necessary equipment can be mail-ordered. An undergraduate at MIT, working alone and using only public information, produced a plausible bomb design in only five weeks.

Even if fissile materials could not be diverted, the operation of a nuclear fuel cycle affords terrorists exceptional opportunities.[29] In November of 1972, three men with guns and grenades hijacked a Southern Airlines DC-9 and threatened to crash it into a reactor at the Oak Ridge National Laboratory if their ransom demands were not met. In March of 1973, Argentinian guerrillas seized control of a reactor under construction, painted its walls with political slogans, and departed carrying the guards' weapons.

A former official in the U.S. Navy underwater demolition program testified before Congress that he ". . . could pick three to five ex–underwater demolition Marine reconnaissance or Green Beret men at random and sabotage virtually any nuclear reactor in the country. . . . The amount of radioactivity released could be of catastrophic proportions."

One visitor to the San Onofre reactor in California recently pulled a knife marked "lethal weapon" and a bottle of vitamin pills marked "nitroglycerine" from his pocket when his tour was next to the control room, to demonstrate how easily the reactor could be penetrated. Various magazine articles have described how a saboteur might initiate a core meltdown in a reactor.

Werner Twardzik, a parliamentary representative in West Germany, joined a tour of the 1,200-megawatt Bilbis-A reactor carrying a 60-centimeter (2-foot) bazooka under his jacket. He toured the world's largest operating reactor with the weapon undetected and presented the bazooka to the power plant's director when the tour ended.

Threats to destroy a reactor in such a way as to release much of the radiation in its core numb the mind. Yet two French reactors were bombed by terrorists in 1975, and several other facilities were bombed in 1976. Between 1969 and 1976, ninety-nine separate incidents of threatened or attempted violence against licensed nuclear facilities were reported in the United States alone. A nearly completed nuclear plant in New York was damaged by arson. A pipe bomb was found in the reactor building of the Illinois Institute of Technology. The fuel storage building of the Duke Power facility at Ocone was broken into. Seventy-six additional incidents took place at government atomic facilities.

If the radioactive iodine in a single light water reactor (LWR) were uniformly distributed, it could contaminate the atmosphere over the lower forty-eight United States at eight times the maximum permissible concentration to an altitude of about ten kilometers (six miles). The same reactor contains enough strontium 90 to contaminate all the streams and rivers in the United States to twelve times the maximum permissible concentration. These materials could not be distributed so uniformly, but the figures serve to indicate that every reactor holds the perils of Pandora's box.[30]

A large fuel reprocessing plant, in addition to being a handy source of plutonium, would contain up to 500 times as much radioactive strontium as a reactor holds. If such concentrated and vulnerable sources of radioactive material became the target of a nuclear explosive—delivered by either a terrorist group or a hostile power—the deadliness of the resulting hybrid would be formidable.

In addition to the perils inherent in the physically discrete stages of the nuclear fuel cycle, problems surround the transport of potentially dangerous materials from stage to stage. Today such transportation is frequently global in scope—witness the British agreement to reprocess 4,000 metric tons of Japanese fuel. In 1974, in the United States alone, 1,532 shipments involving about 50,000 pounds of enriched uranium and 372 shipments totaling about 1,600 pounds of plutonium were made. The record of transportation foul-ups is legendary, and the future danger from either accidental or willful mishaps is commensurate. Moreover, the security accorded even plutonium and highly enriched uranium has been unpardonably lax.

In the general transport of non-nuclear goods, a loss rate of about 1 percent is common. A 1 percent loss of bomb-grade materials could jeopardize world stability; 1 percent of the cumulative expected plutonium flow through the year 2020 would be enough for 400,000 small bombs. Improvements are being made—including blast-off wheels to incapacitate trucks in case of hijackings, and heavy containers that are difficult both to steal intact and to break open. To prevent diversion by skyjacking, some nations have decreed that no airplane may carry enough fissile materials to create a bomb. Even today, however, international shipments of bomb-grade materials and nuclear wastes generally travel unguarded and are subject to accidents or sabotage.

In time, the volume of transportation may be reduced through greater regionalization. The construction of huge self-contained nuclear parks, each housing twenty or more reactors, has even been suggested. In such parks, the entire nuclear fuel cycle could be contained within well-guarded boundaries. Although this setup would reduce transportation problems, it would do so at a high price in terms of both the vulnerability of such centralized facilities and their environmental impact.

Guarding against terrorism requires impossible foresight. Who in 1975 expected a group of South Moluccan extremists to hijack a train in the Netherlands in order to bargain for the independence of the Moluccan Islands from Indonesia? Protecting ourselves against future terrorism means nothing less than building a nuclear system able to withstand the tactics of future terrorists fighting for a cause that has not yet been born.

Nuclear Power and Society

The increased deployment of nuclear power facilities must lead society toward authoritarianism. Indeed, safe reliance upon nuclear power as the principal source of energy may be possible only in a totalitarian state. Nobel Prize–winning physicist Hannes Alfven has described the requirements of a stable nuclear state in striking terms:

Fission energy is safe only if a number of critical devices work as they should, if a number of people in key positions follow all of their instructions, if there is no sabotage, no hijacking of transports, if no reactor fuel processing plant or waste repository anywhere in the world is situated in a region of riots or guerrilla activity, and no revolution or war—even a "conventional" one—takes place in these regions. The enormous quantities of extremely dangerous material must not get into the hands of ignorant people or desperados. No acts of God can be permitted.

The existence of highly centralized facilities and their frail transmission tendrils will foster a garrison mentality in those responsible for their security. Such systems are vulnerable to sabotage, and a coordinated attack on a large facility could immobilize even a large country, since storing substantial amounts of "reserve" electricity is so difficult.

The peacetime risks would be multiplied in times of war. With the proliferation of nuclear power facilities, risks that were previously restricted to atomic arms accrue to conventional weapons. Dr. Sigvard Eklund, director-general of the International Atomic Energy Agency, described the situation to the Swedish Academy of Sciences in 1973:

I emphasize that the maintenance of peace is a condition *sine qua non* for the widespread use of nuclear power which is foreseen. A situation where power reactors above ground would be the object of warfare from the air would have unthinkable consequences, as would, for that matter, fighting action among some of the 100-odd warships propelled by nuclear power.

Nuclear power is viable only under conditions of absolute stability. The nuclear option requires guaranteed quiescence—internationally and in perpetuity. Widespread surveillance and police infiltration of all dissident organizations will become social imperatives, as will deployment of a paramilitary nuclear police force to safeguard every facet of the massive and labyrinthine fissile fuel cycle.

Widespread nuclear development could, of course, be attempted with precautions no more elaborate or oppressive than those that have characterized nuclear efforts to date. But such a course would assure an eventual nuclear tragedy, after which public opinion would demand authoritarian measures of great severity. Orwellian abrogations of civil liberties might be imposed if they were deemed necessary to prevent nuclear terrorism.

The capital-intensive nature of nuclear development will foreclose other options.[31] As governments channel streams of capital into directions in which they would not naturally flow, investment opportunities in industry, agriculture, transportation, and housing—not to mention those investments in more energy-efficient technologies and alternative energy sources—will be bypassed.

With much of its capital tied up in nuclear investments, a nation will have no option but to continue to use this power source, come what may. Already, it has become extremely difficult for many countries to turn away from their nuclear commitments. If current nuclear projections hold true for the next few years, it will be too late. Falsified reports have been filed by nuclear-powered utilities seeking to avoid expensive shutdowns. When vast sums are tied up in initial capital investments, every idle moment is extremely costly. After some level of investment, the abandonment of a technology becomes unthinkable.

In a world where money equals power, large investments in nuclear technology will cause inordinate power to accrue to the managers of nuclear energy. These managers will be a highly trained, remote technocratic elite who make decisions for an alienated society on technical grounds beyond the public ken. They will test C. S. Lewis's contention that "what we call Man's power over Nature turns out to be a power exercised by some men over others with nature as its instrument." As nations grow increasingly reliant upon exotic technologies, the authority of the technological bureaucracies will necessarily become more complete. Some energy planners now project that by the year 2000 most countries will be building the equivalent of their total 1975 energy facilities *every three years*. Although central planners may have no difficulty locating such a mass of energy facilities on their maps, they will face tremendous difficulties siting them in the actual countryside of a democratic state.

A nuclear world would lead to increased technological dependence among nations, especially as the nuclear superpowers conspire to keep secret the details of the fuel cycle. Worldwide dependence upon nuclear power could lead to a new form of technological colonialism, with most key nuclear personnel being drawn from the technically advanced countries. The enormous costs of reactors will result in a major flow of money from poor countries to rich ones.

As the finite remaining supply of petroleum fuels continues to shrink, the need for a fundamental transition grows ever more urgent. The nuclear Siren is at present attracting much interest, but it is to be hoped that her appeal will prove short-lived. Vigorous conservation efforts accompanied by a heroic commitment to the development of benign, renewable resources would be a more judicious course.

It is already too late to avoid widespread dissemination of the engineering details underlying nuclear power. What can still be sought, however, is the international renunciation of this technology and all the grave threats it entails. Although the nuclear debate has been dominated

II

An Energy-Efficient World

4. The Case for Conservation

DOLLAR FOR DOLLAR, investments in increasing the energy efficiency of buildings, industries, and the transportation system will save more energy than expenditures on new energy facilities will produce. This applies to both rich lands and poor. Continued growth in per capita fuel consumption can only imperil the developed world, and "anticipatory conservation" should be a keystone of Third World development. Ironically, the fossil fuels we now devour at an astonishing rate are composed of the leftover food of that prime example of immoderate growth—the dinosaur. Rather than learning from history's mistakes, we are burning the evidence.

Most countries assume that their fuel requirements will continue to grow for the foreseeable future.[1] If the need for an eventual energy ceiling is admitted, the day of reckoning is always thought to lie beyond the horizons of official projections. In chart form, the expected growth in fuel requirements is frequently depicted as an expanding wedge, still winging exponentially skyward in the last year of the forecast.

Such studies, and there have been scores, do not cap an in-depth examination of a spectrum of alternative policies. They make no attempt to grapple with the question "What can be?" They ask only "Where do we seem to be heading?"[2] Projections are judgments made today about tomorrow using data generated yesterday. If the smooth flow from yesterday to tomorrow is disrupted, the projection will prove erroneous. Economist Thomas Schelling has identified this problem as "a tendency in our planning to confuse the unfamiliar with the improbable." Schelling says that "the contingency we have not considered looks strange;

what looks strange is thought improbable; what is improbable need not be considered seriously." An Arab oil boycott, for example, was considered too unlikely to warrant a place in anyone's calculations until history made it a fact.

Because fuel supplies have been fairly flexible, past predictions tended to be self-fulfilling. A high level of demand was forecast; the necessary power plants and refineries were built to meet the posited demand; the fuel and electricity were consequently made available; and the forecast was borne out. Current forecasts, however, have cantilevered such enormous projections of future usage off such small factual bases that the ceilings must eventually topple. To meet these projected levels of demand, thousands of nuclear reactors, countless miles of strip mines, and a large fraction of all available capital would be required. The inevitability of such projections coming true has, therefore, been met with increasing skepticism. Most official forecasts continue to claim that twice as much fuel will be "needed" fifteen years hence as is used today. But more and more people are beginning to ask: Needed for what?

Energy consumption and human well-being do not go hand in hand like Jack and Jill.[3] This common misconception is based upon a presumed relationship between fuel consumption and Gross National Product, and it suffers from three faults. First, the GNP has been largely discredited as a measure of social welfare; second, fuel consumption is a woefully inadequate index of energy use; and, finally, the relationship between GNP and fuel use is remarkably variable among countries and over time.

The GNP—the quantity of goods and services produced and exchanged in the marketplace—is widely accepted as an economic indicator. It is the measure of national economic growth in Nepal as well as in West Germany. However, it provides only partial insight into the well-being of a society. The GNP is a strange agglomerate of goods and evils, of services and disservices—all of which have nothing in common except that they cause money to change hands. The GNP measures with the same inhuman eye the costs of school systems and the costs of prisons for those the schools fail, the costs of nuclear weapons and the costs of diplomatic efforts to persuade people not to use them. The GNP is not reduced by the terrorist bombing of a crowded airport, but it grows as the bodies are buried or mended and the bricks reassembled. It does not shrink along with unique ecological habitats or non-renewable resources, or pale as pollutant wastes are disgorged into the public air and water. The GNP provides no indication of how goods and services are distributed—probably the single most important dimension of social welfare. Nor can a GNP reflect the vital signs of a nation: the pulse of its institutions, the wisdom of its public servants, the strength of its families, the freedom and happiness of its people. In Herman Daly's phrase, the GNP measures "only what can be counted, not what counts."[4]

Just as GNP ignores the qualitative dimensions of life, fuel consumption statistics exclude important qualitative aspects of energy transactions. Discussions of energy requirements in terms such as "barrels of oil-equivalent" can be misleading because, while fuel is consumed, energy—so the First Law of Thermodynamics says—is not. Energy is merely used to perform work. After being used, it still exists. After a unit of fuel has been consumed, the energy it contained takes another form (e.g., electricity, light, motion, or heat). However, use itself does render energy somewhat less useful.[5]

As energy is used, it degenerates into lower-grade heat. Television sets get hot; light bulbs get hot; automobile engines and tires get hot. Heat flows from warmer to cooler objects in a relentless pursuit of equilibrium, becoming ever more dilute and disorganized. As physicists say, its entropy increases. This inexorable increase in entropy is the crux of the Second Law of Thermodynamics. The Second Law thus explains why a given quantity of concentrated, high-quality (low-entropy) energy is more useful for some types of work than is an equal amount of low-quality energy.

Most studies of energy use deal only with its quantitative dimension. They consider the flow of Btu's (or calories or joules) used in a given process, but they do not distinguish among relative entropy levels. They thus ignore the most important aspect of the energy flows they analyze.

Even if one valued the purely quantitative notions of fuel consumption and GNP as analytical tools, the relationship between the two is too ambiguous to be used in policymaking. The amount of fuel needed to produce one dollar's worth of GNP varies by a factor of more than 100, depending upon what good or service is being produced.[6] Energy itself—electricity, oil, and gas—is obviously the most energy-intensive

of goods, followed by products such as cement, aluminum, and miscellaneous chemicals. Medical services and mechanical repairs, on the other
hand, require relatively little energy for each dollar spent. Energy-inten-
siveness varies with both the mix of goods and services in a country's
GNP and the efficiency with which that mix is produced. Sweden and
West Germany, with about the same GNP per capita as the United
States and Canada, use about half as much fuel per capita.[7]

From the end of World War II until 1974, the amount of fuel
consumed per unit of GNP has generally decreased in the industrialized
world, even though the real cost of fuel declined. Technological innovations and shifts in the kind of outputs comprising the GNP account in
large part for this trend. In 1920, fully 141,000 Btu's were needed per
dollar of GNP in the United States. But by 1973 only 89,000 Btu's were
associated with each dollar of GNP. The ratio of fuel use to GNP could,
concludes economist John Meyer in a study for the Conference Board,
continue to fall by 2 percent per year without injuring the economy.[8]
The Energy Policy Project of the Ford Foundation contends that if U.S.
fuel consumption were to level off in 1985, the GNP in the year 2000
could still be within 4 percent of what it would be if fuel use grew at
its historic rate.

Energy is just one of many largely interchangeable factors that
contribute to economic production. Much of the recent exponential rise
in fuel consumption was caused by cheap fuel being substituted for labor
or materials. Fuel use can be cut substantially, without affecting the
GNP, if only this substitution is reversed.

Like certain vitamins, energy is invaluable to a point, sometimes
neutral in its effects after that point has been reached, and actually
harmful in large quantities. Eventually, such hidden costs as environmental deterioration, resource exhaustion, and structural unemployment begin to heavily outweigh the marginal benefits.

Energy and Equity

In 1931, John Maynard Keynes followed a long tradition among
economists—a tradition that encompassed both Mill and Marx—of
distinguishing between those economic products that are truly needed
and those that are merely desired:

Now it is true that the needs of human beings may seem to be insatiable. But they fall into two classes—those needs which are absolute in the sense that we feel them whatever the situation of our fellow human beings may be, and those which are relative in the sense that we feel them only if their satisfaction lifts us above, makes us feel superior to, our fellows. Needs of the second class, those which satisfy the desire for superiority, may indeed be insatiable; for the higher the general level, the higher still are they. But this is not so true of the absolute needs—a point may soon be reached, much sooner perhaps than we are all of us aware of, when those needs are satisfied in the sense that we prefer to devote our further energies to non-economic purposes.

Perhaps two billion people around the world are still striving to meet Keynes' first category of needs. Satisfying the absolute needs of all should be the first order of business in a humane and just world. Fortunately, to the extent that these absolute needs require energy, it can be readily provided from easily tapped natural flows.

Above this level, poverty is a matter of wants rather than needs, of spirit rather than body. This is not to say that this kind is less legitimate or less important to people—merely that it is distinguishable. Persons suffering a poverty of wants are "poor" only in comparison with others who are "rich." If someone earns $5,000 and everybody else on the street earns $50,000, that person is poor. But if someone earns $5,000 and everybody else in the neighborhood (or city or nation) earns only $500, that person is rich. Thus, any legitimate "cure" for poverty will have to alter the *relative* distribution of income and wealth.

It is often held that growth will make redistribution painless. During his Great Society days, President Lyndon Johnson once told his cabinet, "Boys, there's going to be enough for everybody, and that means the folks we have to take a little from won't miss it so much." Yet during this period when fuel consumption and almost any other material indicator signaled enormous growth, precious little income or wealth changed hands in the United States. Consequently, the absolute gap between rich and poor—measured in deflated dollars—grew larger.

A handful of countries, chiefly European, have used the fruits of growth to advance the relative well-being of the disadvantaged. However, none has had the distributional success of China, which had little or no per capita growth during its period of leveling. In most countries, the wealthy prosper most during periods of growth. In agrarian countries the poor often find themselves worse off in *absolute* terms during periods of rapid national economic growth. If poverty is the enemy, only political weapons can fell it: confiscatory inheritance taxes, universal floors and ceilings on income, and other social and economic levelers.

Growth as an Institutional Force

Within some limits, a commercial enterprise can be adjusted to achieve any or several different goals: it can maximize profits, employment, output, or security. The energy industries have largely sought to maximize growth, often at the expense of other objectives. To encourage growth, rates and prices have been structured in ways that reward high consumption. They have conveniently ignored most environmental and health costs.

From the viewpoint of the energy producer, investments in growth have a substantial advantage over investments in conservation: new facilities produce a tangible, salable product. Although the same amount of money invested in conserving energy would often save more energy than can be produced by investments in new facilities, this conserved energy (which would otherwise be wasted) is energy that has already been counted by the producer as sold. The energy company and its stockholders, for whom a dollar burned is a dollar earned, are generally unenthusiastic about "returned merchandise."

The understandable drive to sell increasing amounts of energy has unfortunate consequences. For example, electric utilities have no incentive to match energy types with appropriate uses. Because they sell only electricity, electricity is hawked for all uses. Utilities first encouraged extravagant consumption for appropriate uses of electricity (e.g., lighting). Later, as the "live better electrically" campaign took hold, they couldn't resist pushing inappropriate uses (e.g., space heating) as well.

For most artificial lighting, no better energy source than electricity exists. But artificial lighting itself often becomes too much of a good thing. Lighting requirements were minimal until the industry lobbied tirelessly to shed more and more light on things. William Lam, a Massachusetts architect and lighting consultant, has described how lighting standards for U.S. schools rose from three foot-candles in 1910, to eighteen by 1930, to thirty by 1950, to between seventy and 150 today.

Similar increases took place in office buildings, hospitals, and other public buildings.

Lights give off more heat than illumination. The most efficient fluorescent lamps convert only about 20 percent of the electricity they use into light, casting off the remainder directly as heat. And incandescent bulbs are only about one-third as efficient as fluorescent ones. By the late 1950s, so much heat was being generated by the lights in some commercial buildings that air conditioning was needed even in winter. The sales manager of the Georgia Power Company has explained why this phenomenon warms his heart along with buildings:

. . . if we can get the heating, the other loads come rather easily. If we sell high level lighting, we've got the heating. We also have a much bigger air conditioning load than we otherwise have had. We also have a high load factor heating system that operates all year long! The air conditioning will operate all year long! [The current lighting standards] will get you the totally electric job. . . . It is the inside track, the sure thing we have been looking for.

Fuel shortages, environmental constraints, political opposition, and a growing unwillingness to commit most of their discretionary capital to the construction of new energy facilities have forced many nations to question whether burgeoning Btu consumption is in their best interest. In virtually every country the search has begun for comprehensive energy-conservation strategies.

A society intent upon reducing its fuel consumption can turn to both technical solutions and social solutions. Technical solutions require essentially no behavioral alterations—merely changes in the types of machinery we utilize, or in the way we use it. Social solutions, on the other hand, require changes in the way people live and act.

Technical Approaches[9]

Two basic kinds of technical approaches are leak plugging and machine switching. Leak plugging eliminates the waste in existing technologies, while machine switching involves the replacement of existing devices with more efficient ones. To insulate a house is to plug a leak; to replace an electrical resistance furnace with a heat pump is to switch machines. To tune up a car is to plug a leak; to trade it in for a more fuel-efficient model is to switch machines.

A less obvious kind of technical solution involves the careful thermo-dynamic matching of the task at hand with the energy sources best able to perform it without generating waste. Initial "compatibility" studies in several countries have uncovered enormous inefficiencies; high-grade useful energy is habitually treated as a waste product and discharged into the environment. A group of physicists who scrutinized the efficiency of U.S. energy use in terms of the Second Law of Thermodynamics for the American Physical Society pegged the country's over-all thermody-namic efficiency at between 10 and 15 percent.[10] Cars were found to be 10 percent efficient, home heating 6 percent, air conditioning 5 percent, and water heating only 3 percent efficient.

A thermodynamic efficiency of 100 percent is an idealized and impossible standard. Moreover, decisions cannot be made on the basis of thermodynamic efficiency alone; economic costs, environmental costs, and the costs of human time must all be balanced in a wise strategy. Nonetheless, an efficiency as low as 10 to 15 percent should raise eyebrows. Doubling it to a mere 20 or 30 percent would cut the U.S. energy budget in half without changing anything other than the usefulness of machines and processes, and recent studies confirm that such a move is practical.[11]

Every country uses most of its energy as heat. In many, heat com-prises over 90 percent of energy demand, while in the United States the figure ranges closer to 60 percent. In industrialized countries, much of this heat is obtained by burning fossil fuels at more than 1,000 degrees Centigrade—often to heat water or air to less than 100 degrees C. Even worse, these fuels are often converted at 40 percent efficiency or less into electricity, which, after transmission and distribution losses, is used in domestic hot-water heaters. Using electricity to heat water is akin to killing houseflies with a cannon; it can be done, but only with a lot of messy, expensive, and unnecessary side effects. It would be much more thermodynamically efficient to reserve the high-temperature heat and electricity for tasks that require them, and to use residual heat for lower-grade purposes, like heating water. Alternatively, low-grade heat could be pumped from another source and upgraded just the last few degrees by burning fossil fuels.

Finally, finite fuels can be replaced by sustainable energy sources, drawing upon the natural flows of energy that will circulate through the biosphere whether or not they are tapped by human beings. At present, we tend to ignore the sun and the wind as power sources, or to use our fossil fuels to resist their effects. Instead, we could harness them to meet human energy requirements.

Probably the strongest single impetus for technical approaches to conservation has been economic. In both industrialized and rural societies, a dollar invested in energy conservation can make more net energy available than a dollar invested in developing new energy sources. Eric Hirst calculates, for example, that investments in improving air conditioner efficiency can save ten times as much electricity as similar investments in new power plants can produce. Arjun Makhijani has shown how a $10 investment in improved stove efficiency can cut an Indian family's wood consumption in half—saving $10 to $25 per year. Neither example entails a loss of benefit or comfort. Both save far more energy per dollar than investments in new energy sources could produce, and the energy saved is just as valuable as new energy produced.[12]

The economic advantage of such conservation speaks for itself, especially in a period of general capital shortages. Roger Sant, former assistant administrator of the U.S. Federal Energy Administration, has argued that a $500 billion investment in energy conservation would save the United States twice as much energy as a comparable investment in new supplies could produce. Of course, every society has large investments sunk in existing buildings and machinery, and sizable savings can be achieved through conservation only gradually, as existing capital is replaced by newer, more efficient items. But such investments should not blind us to the advantages of beginning the gradual changeover to wise energy management now.

Social Approaches

The most elementary of the "social" approaches to energy conservation might be thought of as belt tightening. This conservation tactic generally refers to minor changes in life style that are mostly neutral in their effect on people but that are occasionally inconvenient or irritating. Belt tightening involves, for example, such things as turning off unnecessary lights, driving cars more slowly, and using commercial or residential heating and cooling systems more sparingly.

Social approaches might also include cooperative endeavors: car pools, public transit systems, apartment buildings, joint ownership or rental of infrequently used items, and so on. A four-person car pool uses only about one-fourth as much gasoline as do four cars driving the same distance, and most apartment house walls, since they are shared, retard heat loss to the outdoors.

The final social approach to energy conservation involves exchanging energy-intensive devices for those that require less energy. The evolution of living habits is already evident in the general shift of most industrial societies from an emphasis on goods to an emphasis on services. It could lead to the substitution of low-energy activities like gardening or education for high-energy activities like skydiving. Their proponents frequently call low-energy life styles ways of "living lightly on the earth." Undertaken by entire societies, such social changes could cut fuel consumption down to size by reshuffling the components of the GNP.

The Politics of Conservation

The case for conservation is compelling. This does not, however, mean that effective programs will inevitably or even probably take shape.[13] In fact, in a report entitled "Energy to the Year 1985," the Chase Manhattan Bank claims that there is no scope for conservation whatsoever, even in the United States.

It has been recommended in some quarters that the United States should curb its use of energy as a means of alleviating the shortage of supply. However, an analysis of the uses of energy reveals little scope for major reductions without harm to the nation's economy and its standard of living. The great bulk of the energy is utilized for essential purposes—as much as two-thirds is for business related reasons. And most of the remaining third serves essential private needs. Conceivably, the use of energy for such recreational purposes as vacation travel and the viewing of television might be reduced—but not without widespread economic and political repercussions. There are some minor uses of energy that could be regarded as strictly non-essential—but their elimination would not permit any significant savings.

This statement, and others like it made by the energy industry and its financial backers, simply ignores the physical and technical phenomena of the world around us. Because those who draft such reports assume an efficient marketplace has eliminated all waste, they fail to note leaky buildings, inefficient machinery, and workers' disinclination to save money for management. They also ignore the fact that credit criteria systematically channel capital to big projects (like power plants) rather than to small ones (like home insulation)—even when the small ones would bring a higher energy yield.

The fundamental flaw in the Chase statement is that it confuses energy conservation with curtailment. Curtailment means a cold house; conservation means a well-insulated house with an efficient heating system. Curtailment means giving up automobiles; conservation means trading in a seven-mile-per-gallon status symbol for a forty-mile-per-gallon commuter vehicle. Energy conservation does not require the curtailment of vital services; it merely requires the curtailment of energy waste.

Recent economic history, especially in the industrialized world, has been molded by Chase-style thinking. And the past is widely presumed to be prologue to the future. This presumption guides the elaborate computations of most modern forecasting, and it underpins much of our conventional wisdom. But, as René Dubos has written, "Trend is not destiny." Calamities and booms can intrude upon the smooth curves of extrapolation; people and nations can rethink their direction and alter course.

For students of energy policy, the future is not what it used to be. Consumption patterns for commercial fuels, after two decades of unbroken exponential growth, have changed radically over the last two years. Even more fundamental discontinuities seem likely to appear in the near future. Momentous conflicts loom between habits and prices, between convenience and vulnerability, between the broad public good and narrow private interests.

A comprehensive program of energy conservation initiated today will allow the earth's limited resource base of high-quality fuel to be stretched. It will enable our descendants to share in the earth's finite stock of fossil fuels. It will make an especially critical difference to those living in underdeveloped lands where the marginal benefit per unit of fuel used is far greater than it is in highly industrialized countries.

Energy conservation will allow a portion of the fossil fuel base to be reserved for non-energy purposes: drugs, lubricants, and other materials.

The energy cost of manufacturing such substances from carbon and hydrogen when our existing feedstocks have been exhausted will be astronomical.

Energy conservation will allow us to minimize the environmental degradation associated with all current energy conversion technologies. It will decrease the odds that we will cross climatic thresholds, triggering consequences that may be devastating. It will provide the opportunity to avoid reliance upon objectionable energy sources while the search for safe, sustainable sources continues.

Energy conservation could lead to more exercise, better diets, less pollution, and other indirect benefits to human health. An enlightened program of energy conservation will substantially bolster employment levels. And the security of a modest energy budget is more easily assured than that of an enormous one that depends upon a far-flung network of sources.

Recognizing that circumstances have changed fundamentally, the world can undergo the transition into a new era without tumultuous upheavals. But should we fail to come to grips with the new energy status quo now, the world may permanently forfeit that chance. The newly recognized potential for energy conservation is a challenge and an opportunity. In the past, conservation was viewed as a marginal activity of do-gooders. Today, saved energy is the world's most promising energy source.

5. Watts for Dinner: Food and Fuel

THE AVERAGE WELL-FED person uses about as much energy each day as a steadily glowing 100-watt bulb. This energy, measured as the calories in food, is in fact stored solar energy. Like all other animals, *Homo sapiens* cannot capture sunlight directly and must depend upon plants to gather radiant energy and to make it "edible."

Through photosynthesis, plants convert sunshine into chemical energy. Using only about one-sixth of the energy it captures to sustain itself, a plant stores the remaining five-sixths in chemical bonds. Sooner or later, these bonds are broken by animal metabolism, fire, or the slow processes of decay.

Human beings cannot use all the energy available in the chemical bonds of plants. For example, less than half the dry mass of a corn plant is grain. Most of the energy in the portion that can be digested is not retained by human beings either; most passes through and remains stored in excrement. About 20 percent of the potential energy in digestible food is all human beings usually retain.

The sunlight plants capture works its way through the animal kingdom along food chains, losing energy at each level. The longer the chain, the lower the percentage of original energy available at its terminus. The energy losses along one such food chain have been described by Lamont Cole:

For example, 1,000 calories stored up in algae in Cayuga Lake can be converted into protoplasm amounting to 150 calories by small aquatic animals. Smelt eating these animals produce 30 calories of protoplasm from the 150. If a man eats the smelt, he can synthesize six calories of fat or muscle from the 30. If he waits for the smelt to be eaten by a trout and then eats the trout, the yield shrinks to 1.2 calories.

Human beings stand at the top of many food chains. We eat other animal and vegetable species, but are rarely ingested ourselves. As we grow more prosperous, we tend to select the components of our diet from farther up the food chain. As we ascend, the energy indirectly contained in our diet rises as well. Postwar Japan saw a great rise in meat consumption; a comparable phenomenon now appears to be emerging in some oil-exporting countries. Similarly, per capita intake of beef in the United States has more than doubled in recent decades. As a general rule, the wealthier the country, the more energy its typical diet contains.

In terms of energy efficiency, the history of agriculture has been a story of near constant decline. Hunting and gathering societies, as anthropologist Marshall Sahlins observes, invest less energy in obtaining a unit of food than do societies with planned cultivation. Indeed, domesticated food crops can become so dependent on human intervention that some cannot even disperse their own seed or compete in a natural ecosystem. Such crops require planting, cultivation, fertilization, and irrigation.

In the early days of agriculture, the energy put into cultivation was all derived from human muscle. Human beings, in turn, culled all their energy from food. Unless the agricultural system had produced more food energy than was expended in muscle power to grow the crops, agriculture would have perished, and with it the first farmers. No creature can persistently spend more bodily energy to acquire its food than it derives from that food; it must at least break even.

To the extent that a foraging animal, or a fuel-driven engine, was substituted for muscular energy, the ratio of human energy invested to food calories acquired diminished. At the same time, the ratio of total energy invested to food calories acquired swelled. But since people could not eat grass or oil, and since both seemed to be plentiful, total energy accounting was not, until recently, given serious attention by farmers. Ratios of food production to units of land, labor, fertilizer, or seeds were often noted, since these factors obviously limited production. But fuel was not considered a limiting factor and food production increases were generally achieved through the use of additional fuel. Today, Professor David Pimental of Cornell University calculates, the average U.S. farmer uses the energy equivalent of 80 gallons of gasoline to raise an acre of corn.[1]

As people moved off farms into cities, food had to be stored for longer periods and transported farther. For example, as America became increasingly a leisure society, the popularity of food became directly related to its ease of preparation. Today, a vast food infrastructure, built in large measure around the food processing industry, delivers more than three-fourths of U.S. food pre-washed, pre-cooked, or otherwise prepared. The corporate kitchen has taken over many tasks traditionally performed in the home—substituting fuel and machinery for human labor.

Farming now accounts for less than one-fifth of the total energy use in the American food system. The remaining four-fifths are used to process, distribute, and prepare the food.[2] Almost twice as much energy is used to process food (33 percent) as to grow it (18 percent). Another 30 percent of U.S. food-related energy is used for stoves, refrigerators, trips to and from the supermarket. Wholesaling and retailing use 16 percent, while commercial transportation accounts for 3 percent. In industrial countries, by far the greatest savings are to be made in food processing and marketing, and in household preparation. However, the farm also holds great scope for increased efficiency and for increased reliance on sustainable energy sources.

Farm Energy

Two twentieth-century phenomena greatly expanded world food demand: population growth and rising prosperity. Of the 30-million-ton annual increment in world grain consumption in recent years, 22 million tons is swallowed up by population growth, while 8 million tons reflects rising affluence. Roughly one-third of the world grain harvest is now channeled into feedlots to fatten cattle, even though feedlot beef has more saturated fats and less protein than grass-fed beef.[3]

Coping with outbreaks of starvation in the developing world has become a principal focus of global humanitarian efforts; in 1966 and 1967, for example, more than one-fifth of the entire U.S. wheat crop was shipped to India to ward off famine. Such efforts—necessary in a crisis, but unsatisfactory in the long run—seem particularly superficial since virtually every country in the world has the physical resources to provide its present population with an adequate diet. Serving the goal of a hunger-free world, in part by initiating necessary land reforms, would seem a natural and popular course for governments to pursue. However, the record is dismal. Although every continent except Western Europe produced a net food surplus in the 1930s, continent after continent fell into food deficit over the next forty years. Only North America and Australia have surpluses today.[4]

Grain farmers in North America and Australia produce as much as they do in part because they use so much fuel. North America and Australia both use several times more energy to produce, process, retail, and prepare the food they grow than the food itself contains. Yet none of the energy in the fuel is actually transferred to the food. Fuels used in the food system merely substitute for labor, land, capital, rain, and so forth—not for the sunshine from which food energy issues. If the entire world ate food grown, processed, and distributed in the American style, the global food system would consume most of the world's total fuel production, leaving little for industry, transportation, or even home heating. Yet most of the world aspires to the American diet, and the techniques used to produce the world's food are becoming ever more energy-intensive.

The problem of feeding the world's hungry has sometimes been misperceived as a technical problem, for which a technical solution is nicely in hand. The last decade has seen a rapid global proliferation of high-yielding varieties of grains (HYVs) and the energy-intensive cultivation methods these varieties require. This agricultural phenomenon—originating in the industrial world, but widely applied in the Third World—is commonly referred to as the Green Revolution.

Taking full advantage of the new miracle grains requires large amounts of energy. High yields can demand chemical fertilizers and pesticides, irrigation equipment, and farm machinery—all energy-intensive to make and use. Transforming traditional agriculture also demands considerable up-front capital, so the primary benefits of increased productivity tend to flow to those with land, money, or political influence.

The Green Revolution originally appeared to many to be a timely answer to widespread hunger in an age of cheap, abundant fuel. Undeniably, it staved off certain starvation for millions of people. But in recent years fuel has been neither cheap nor abundant. Instead, energy shortages have constrained agricultural productivity increasingly. For international agriculture, the implications of this change can scarcely be exaggerated.

Rising demand for food in a world with limited naturally watered fertile cropland is leading farmers everywhere toward energy-intensive changes in their traditional practices. Chemical fertilizers are substituted for land, and irrigation is substituted for rainfall. While energy-efficient practices must be encouraged, and the use of sustainable energy sources promoted, all is futile unless population growth and rising meat consumption can be controlled.

Fertilizer

As virgin agricultural land has grown scarce, farmers have begun to use more and more chemical fertilizers to boost production on existing farms.[5] Since chemical fertilizers—and nitrogen fertilizers in particular—are highly energy-intensive, energy consumption has risen with fertilizer use. U.S. corn farmers, for example, now use more energy per acre in fertilizer (940,800 kilocalories) than in tractor fuel (797,000 kilocalories). Fertilizer prices, unfortunately, have escalated steeply, since they bear the imprint of oil and gas price hikes.

Natural gas, which is used in the manufacture of most nitrogen fertilizer, is plentiful enough at the moment. In fact, the amount of gas flared—that is, wasted—worldwide each year is twice the amount needed to maintain the current world output of nitrogen fertilizer. However, gas production in the continental United States peaked in 1974, and world gas production is expected to peak before the year 2000. The price of natural gas has already begun to climb, reflecting this long-term scarcity.

Responsiveness to large dosages of chemical fertilizers is the premier advantage of high-yielding varieties; without such fertilizers, HYVs yield little more per acre than do traditional crops. Hence, with the spread of high-yielding varieties has come the rapid expansion of chemical fertilizer use. Fertilizer increments would bring the greatest returns in poor countries where little is now used. However, most poor nations cannot easily afford to increase their use of fertilizer at the new high prices. In 1975, American agriculture used about 20 million tons of chemical fertilizer. By comparison, India, with about the same amount of farmland and with two and a half times as many people to feed, used only 3 million tons.

Substitutes for and complements to chemical fertilizers abound. Traditional agricultural practices that were abandoned during the era of cheap energy could, for example, be revived. Some are today making a comeback deep in the U.S. breadbasket. Richard Thompson, who operates a 285-acre midwestern farm without using chemical fertilizers, uses manure from his cows, and sewage sludge from nearby Boone, Iowa, to enrich his land. He also plants and then plows under "green manure" —legumes such as soybeans, alfalfa, and clover, which have nitrogen-fixing bacteria in their root nodules. He carefully rotates his crops in a regular cycle of corn, soybeans, corn again, then oats and hay—a practice that also helps control insects and diseases.

When a team of scientists from Washington University studied fourteen pairs of crop-livestock farms in the U.S. corn belt, it found that over-all production on fourteen organic farms was 10 percent lower than production on fourteen farms that used chemical fertilizers and pesticides. The organic farms required about 12 percent more labor per unit of market value, but only half as much energy as their counterparts. The financial returns were about the same for both groups of farms, largely because of the savings on fertilizers.[6] Many of Richard Thompson's neighbors, for example, invest as much as $80 per acre in chemicals, an annual extra expense of $23,000 for farms the size of Thompson's.

Seemingly newfangled, Richard Thompson's farming practices have two hundred years of "field tests" behind them. By the mid-eighteenth century, Edinburgh, Scotland, was operating a sewage farm, and by mid-nineteenth century extensive sewage farming had begun in Paris and Berlin.[7] In Wassmannsdorf, Germany, a system was devised in 1920 to pipe sewage sludge to farms, using pumps powered by methane produced by the anaerobic digestion of the sewage. Today, Tel Aviv's sewage helps support fruit and vegetable production on the Negev desert. Sewage has long been valued as a fertilizer in several Asian countries, and in China nutrient recycling now approaches maximum efficiency. (In an effort to further boost yields, China has become the world's largest importer of nitrogen fertilizer, and is currently building ten giant fertilizer plants. However, chemical fertilizers always complement—rather than replace—organic fertilizers in China.)

Since nitrogen constitutes 80 percent of the earth's atmosphere, nitrogen shortages pose no threat. The trick is to remove nitrogen from the air in a form that plants can use and that farmers can afford. In natural systems, microorganisms that grow on the roots of some major food plants, including soybeans and alfalfa, perform this task. A Brazilian scientist, Johanna Doebereiner, succeeded in cultivating these organisms on corn roots, a feat recently duplicated at the University of Wisconsin. Such laboratory breakthroughs lead to speculation that corn and other crops might someday satisfy much of their craving for nitrogen without using chemical fertilizer. While not without costs and risks, such an approach could yield large energy savings if it proved successful.

Irrigation

In 1800, 20 million acres of the world were irrigated. Over the next century, the total swelled to about 100 million acres. By 1950, about 260 million acres were irrigated, and by 1970 the total had increased to 470 million acres. The rate of expansion of irrigated land thus actually outpaced the rate of human population growth.

The appeal of irrigation is obvious. Pumped water can allow parched land to be cultivated, can parry the risk of drought, and can boost crop yields. Virtually all crops benefit from a bountiful predictable supply of water, and some of the more productive new crops need water at specific times, making irrigation a necessity.

Where agricultural lands have underground water of reasonable quality, tube wells should replace or complement streams and reservoirs. Tapping the local water table directly, tube wells are not subject to siltation, a process that limits the life spans of dams and reservoirs and that is kept under control in irrigation canals only through extensive maintenance. However, tube wells can be abused. When water is withdrawn from a water table more rapidly than it collects, the table ceases to be a renewable resource. In central Arizona, where industrial and residential users meet farmers at the wellhead, the water table is falling ten to twenty feet a year.

Water is heavy, and lifting it can require prodigious amounts of energy. Electricity use on U.S. farms rose from 15 billion kilowatt-hours in 1950 to 39 billion kilowatt-hours in 1975. About three-fourths of the 1975 total was used for irrigation. Although modern irrigation systems rely mostly upon non-renewable fuel sources for power, they can also be powered with renewable sources. The oldest of these faithful and ever-lasting sources is gravity, which captures rain at higher elevations and tirelessly channels it downhill. Some two-thousand-year-old Roman aqueducts still function admirably without ever having consumed a drop of oil.

China, with about 40 percent of the world's total irrigated land, has also put gravity to work. Four-fifths of its irrigated land depends upon gravity-fed or animal-powered systems. These systems, usually constructed by agricultural laborers during the winter off-season, often lack the capacity to sustain intensive cultivation, but they do protect the land from moderate droughts.

A wide variety of renewable sources can be harnessed to lift water. Simple wind power was the technology of choice until the advent of cheap fuel and electricity. Today, windmills are enjoying a revival in many countries. Traditional windmills are being modified to take advantage of modern aerodynamic theory and to utilize local materials.

Two other power options can be used in conjunction with irrigation systems. Solar pumps, productive on hot days when water demand is highest, are now being used in Mexico, Brazil, and Senegal, though current designs remain economically uncompetitive except for areas exceedingly remote from other power sources. Biogas, a mixture of methane and other gases produced from animal excrement and crop residues, may be a significant new fuel; already some conventional pumps in India use this fuel.

Much more water is delivered to most irrigated fields than is needed to sustain crops. As fresh water becomes scarce in more and more parts of the world, irrigation techniques that use water more efficiently must be devised. One possibility is trickle irrigation, a method in which a small amount of water is delivered in a measured amount to each plant. Trickling is costly, but it saves water and energy and it offers an alternative to the profligate technologies that could well leave the world high and dry.

Farm Machinery

U.S. agriculture prides itself on its enormous productivity per worker. Today, each American farmer feeds fifty of his fellow citizens and, in addition, produces a surplus for export. Only one-tenth of one percent of the world's population works on U.S. farms, but they produce almost one-fifth of the world's grain.

If we consider agricultural labor as the amount of time spent to produce a unit of output, a New York farmer spent 150 minutes producing a bushel of corn in the early twentieth century. In 1955, it took him just 16 minutes. Today, he spends less than 3 minutes per bushel.

Worker productivity grew largely because fossil fuels were substituted for human labor. In the United States, this development—at least in its early stages—was fortuitous. Mechanized farming reduced the need for agricultural labor at the same time that industry required an expanded work force. Between 1920 and 1950, the proportion of the population involved in agriculture decreased by half. In 1962, it halved again. Now it has shrunk by almost half again, and more than 50 percent of the remaining farmers hold second jobs off the farm.[8]

When Great Plains farmers traded in their draft animals for tractors, they no doubt made a wise move. But the introduction of large-scale mechanized farming in poor countries today can be economically inefficient and socially disruptive. When the peculiar needs and conditions of the recipients are ignored in a technology transfer, the "solutions" the new technology produces may prove more troublesome than the problems it was supposed to solve. In those many countries in which 80 percent of the work force is engaged in agriculture, the objective must not be to make every employed laborer as productive as possible, but rather to make the most productive possible use of the entire labor pool.

The substitution of fuels and machines for labor in poor countries has been frequently and understandably condemned. However, the situation is more complex than many critics have acknowledged. Because agriculture is a cyclical activity, the demand for labor ebbs and flows throughout the year. Rice transplantation and grain harvesting, as just two examples, demand an enormous labor pool, but demand it for only relatively brief periods. The wide fluctuations in labor demand can be smoothed out with multiple cropping, which often requires irrigation. But even a series of regular employment peaks will leave the bulk of the labor force underemployed for much of the work year. Even as electrical generating facilities in developed countries build the capacity to meet brief peak periods of demand for electricity, so farmers throughout much of the world raise families large enough to meet their peak labor needs. The careful employment of appropriate technologies to shave some of the labor demands from these peak periods would help to smooth out the employment peaks, increase average labor productivity, and reduce one major impetus to continued population growth. In many cases, such technologies would also increase over-all food output.[9]

The suitability of a particular technology can be measured by its impact on a culture. Accordingly, the purchase of sophisticated equipment may represent a misuse of scarce capital in developing countries. However, use of such devices as the "walking tractors" or two-wheel power tillers common in Japan and Taiwan, and a new Chinese invention for mechanically transplanting rice, may benefit an entire society.

As we approach the end of the petroleum era, the designers and users of farm equipment must accord fuel efficiency a high priority, even as they begin the transition to the use of alternative energy sources. Farming, more than any other commercial activity, has the capacity to become largely energy self-sufficient. The sooner the groundwork is laid for agricultural fuel conservation, the more oil and gas will remain for other purposes.

Crop Drying

A final major use of energy on the farm is grain drying, a technique that permits farmers to minimize field losses by harvesting their crops before they are dry enough to be placed in long-term storage. High-speed grain drying can sometimes use more fuel than tilling, cultivating, and harvesting the grain. Fuel consumption for U.S. tractors and combines generally ranges between five and fifteen gallons per acre. By comparison, reducing the moisture content of 100 bushels of corn from 25 to 15 percent moisture content (from the harvesting stage to the safe-storage stage) in high-speed dryers can require up to twenty gallons of propane fuel. Solar energy can usually be employed for such grain drying, and suitable solar techniques are being developed in many parts of the world.

Drying poses a particular problem in countries attempting to harvest two rice crops during one monsoon. The first crop must be harvested during the heavy rains. Since there is little sun during the monsoon season, grain-drying equipment may have to be powered by methane generated through anaerobic digestion of field residues and other organic matter.

Home Gardens

Home gardens have proliferated throughout Europe and America in recent years. City planners in many parts of the world are now incorporating garden-sized tracts in their designs. In New Bombay, for example, planners hope that each family will raise some fraction of the food they eat. Two Indian journalists reported from China in mid-1976 that wherever they went, they "did not spot even a tiny piece of earth which was not put to use. Gardens attached to houses, even land between telegraph poles and beside the railway track, all of which lie waste in India, were cultivated."

As energy prices, and consequently food prices, soar, more back yards and vacant lots in the industrial world are also being converted into gardens. Many American companies, churches, and schools have set aside plots for private gardens; half of all Americans now grow some of their own vegetables.[10] The English tradition of public land allotments has been revitalized; over half a million gardeners each have a 300-square-yard allotment in Britain, and each plot produces around $300 worth of vegetables a year. Personal greenhouses are also making a rapid comeback in the temperate zone as a means of lengthening the growing season.

A home vegetable garden saves energy in three important ways. First, the gardener's labor (called "recreation") is substituted for gasoline. Second, compost piles provide rich fertilizer while simultaneously reducing the amount of organic residential garbage to be hauled away. Third, growing food at home eliminates much of the need for fuel for processing, packaging, retailing, and transporting the farm-grown commodities. In addition, home gardens require fewer pesticides, partly because crops can be mixed to provide a less attractive target for pests. Home gardens also cut down food waste; people who would not buy a blemished tomato will eat one out of their own garden.

Non-Farm Energy Use

What happens to food *after* it leaves the farm affords the best opportunities for saving energy in the food system. In the industrial world, the food passes through an elaborate infrastructure in which it is inefficiently processed, transported, stored before being prepared and eaten by the consumer.[11] In the Third World, the storage and preparation of food by the consumer entail the greatest inefficiencies.[12]

The food processing industry, like other industries, grew up in an era of cheap fuel prices. As a consequence, it uses energy inefficiently. Most of the energy it consumes is used in the form of low-grade heat, much of which could be provided by elementary solar technologies.

One of the oldest of the food processing technologies is refining. White flour was once universally considered superior to whole wheat flour, as was refined sugar to unrefined sugar. When it was discovered that white flour lacked basic nutrients contained in whole wheat flour, the industry restored some of the lost nutrients to "enriched" flour. Now, however, the evidence is mounting that this enriched flour is still inferior, because the missing fiber content performs a vital health function. Energy is expended refining and then enriching white flour, yet the final product remains in many ways inferior to whole wheat flour.

The food processing industry must also take responsibility for the "fast food" concept. Once food was purchased at a store, taken home, and cooked. Fast foods, however, are cooked at a factory, placed in aluminum trays, sealed with foil, quickly frozen, folded into a paper box, shipped by freezer cars to supermarkets, stored in frozen food bins, driven home, placed in the consumer's freezer, and then eventually cooked again in an oversized, under-insulated oven. The energy used on the food after it leaves the farm is several times greater than that used on the farm.

Food processors must shoulder blame for an explosive growth in unnecessary packaging too, a waste even more pointless than the circular flour "enrichment" process. According to the U.S. Environmental Protection Agency, "The consumption of food in the United States increased by 2.3 percent by weight on a per capita basis between 1963 and 1971. In the same period, the tonnage of food packaging increased by an estimated 33.3 percent per capita, and the number of food packages increased by an estimated 38.8 percent per capita."

Packaging has doubtless reached its apex at the modern American hamburger stand. There, a hamburger comes wrapped in cellophane, surrounded by a circular strip of cardboard, and inside a multicolored cardboard box that is itself placed inside a bag. This tawdry swaddling is usually chucked into a plastic-bag lined garbage can (along with the plastic containers for catsup and mustard, the paper containers for salt and pepper, the paper napkins, and the sales slip) before the hamburger is five minutes old.

Although over-refining, over-processing, and over-packaging should be eliminated, the food processing industry can serve a legitimate function in an urbanized society. But enormous scope exists for improving the energy efficiency with which this function is fulfilled.

Food retailing suffers from some of the same energy inefficiencies that plague other commercial enterprises. Space heating and cooling fixtures are poorly contrived; open entrances and exits are constant drains on space conditioning systems, and so on. Other food retailing problems are unique, including the energy drain of open-topped food freezers, and the strain such freezers place on a store's heating system.

Like "fast food," the supermarket has altered energy tastes and appetites. When neighborhood markets prevailed, trucks delivering food had to make more stops, but the food was then purchased by people who usually carried it home on foot. Now trucks deposit the food at a central supermarket, and hundreds of two-ton private automobiles each transport thirty pounds of food and packaging from supermarket to home.

In the United States, cooking, refrigeration, home freezers, and car trips to the grocery store account for about 30 percent of the total energy expenditures on food—50 percent more than farming does. In fact, more than half the total electricity spent on food is used in homes to power food-related appliances. While some domestic energy use has been transferred to the food processing industry, many frozen foods now require more energy use at home than did their unfrozen predecessors —in addition to the energy used by industry to process them.

Third World families also waste energy on food storage and preparation. Indeed, as a West African saying goes, "It costs as much to heat
the pot as to fill it." Severe shortages of wood for cooking have grown
common in many poorer countries. In much of the Third World, the
wood each person uses for cooking in a year contains between 5 and 7
million Btu's. By comparison, the energy content of the coal burned
annually to generate electricity for a typical electric stove in the industrial world totals about 3 million Btu's, while gas stoves without pilot
lights require only about 1 million Btu's. The widespread use of more
efficient wood stoves could substantially reduce the escalating demand
for wood in the Third World. Biogas stoves could achieve even higher
efficiency, and small, cheap solar cookers need no fuel at all. Solar
cookers are being promoted in more than a dozen countries; one Indian
model, 1.4 meters in diameter, retails for $6.70. Pressure cookers, too,
require much less energy than do standard pots, and cheap, locally
produced pressure cookers could greatly improve fuel efficiency.[13]

The Distribution Problem

The current vogue in some circles is to reduce the food problem to
a single dimension: distribution. There is no question but that distribution is vitally important. The much heralded Chinese agricultural success, for example, may be correctly viewed as primarily a distributional
success. Though the per capita food available in the People's Republic
of China in 1976 was only modestly greater than the amount available
in 1950, virtually no Chinese seem to suffer from hunger and malnutri-
tion. Brazil, with a per capita GNP three times as high as China's,
appears to have far more underfed people than China, especially in the
desperately poor northeast region. Hunger, a sign of extreme poverty,
reflects the inequitable distribution of a nation's wealth as well as over-all
scarcity.

Redistribution of land as much as redistribution of food is necessary
to alleviate global hunger.[14] Land reform will grow even more important
as fossil fuels become more expensive. Small decentralized farms afford
a great many options not available to *latifundia*, or agri-business con-
glomerates. Biggest may have been best in an era of cheap, concentrated
fuels, but a smaller plot holds more advantages in an age of increasing

reliance on such diffuse energy sources as wind and crop residues. Biogas plants and nutrient recycling are most effectively accomplished on a family farm. And the employment that small farms offer can slow or even reverse the mass migration of the rural poor to the cities.

Studies in many countries have found that small, labor-intensive farms tend to produce more per unit of land than do giant farms. In many Third World countries, an increase in the food produced per acre is far more important than an increase in the number of acres a farmer can cultivate. A wise land reform strategy can result in higher total food production as well as more equitable distribution.

In North America, one- and two-person farms large enough to take advantage of mechanization have been found to be as efficient as, and in some instances more efficient than, giant corporate farms. The love the individual farmer has for his land, his capacity for the hard work John Kenneth Galbraith calls "self-exploitation," the intimate knowledge he acquires over decades of living on the soil, are simply not part of daily life on the huge estates that are increasingly dominating world agriculture. Viable small farms are also an attractive alternative to mushrooming urban complexes that depend utterly upon fossil energies.

While redistributing land would help eliminate hunger, it would not, at current production and population levels, improve most diets beyond mere adequacy. Much agricultural production has its roots in delicate environments that cannot long sustain it. Growing populations and declining fossil fuel sources will strain these limits more intensely. If the rural poor are to move from survival to security, if they are to exchange "get-by" meals for well-balanced, interesting diets, if they are to acquire the surpluses and the diversification to ensure that their children will live more comfortably, more than land ownership will have to be reformed. The population explosion must be defused, renewable energy technologies must be widely disseminated, and environmentally sound, sustainable farming practices must be adopted.

Energy is rapidly becoming the most critical variable in the world food system. Little good unused agricultural land remains to be brought under cultivation. Farming marginal lands brings dire, and sometimes irreversible, ecological consequences. Yet the productivity of existing farmlands can be increased only through wise energy use.

Farmers should turn to renewable energy sources for an increasing

6. Energy and Transportation

Each generation travels farther and faster than the previous one. The sheer volume of transport—the movement of people and freight—has swollen many traffic arteries to the bursting point. This growth has been accompanied by a systematic shift of people and goods to less energy-efficient modes of travel. Both trends have been somewhat more dramatic in industrialized countries than in agrarian states, and both are much more pronounced in cities than in rural areas. However, to some degree the general patterns hold throughout the world. The United States, where transportation now accounts—directly and indirectly—for 42 percent of *all* energy use, is leading this trend. Transport fuel alone represents 25 percent of the American energy budget, and an additional 17 percent is used to build and care for vehicles, to construct and maintain roads, and so on.

This heavy commitment of energy resources to transportation is troubling in itself, but the situation looks even more disturbing when the nature of our energy resources is considered. Petroleum, the fuel that existing transportation networks run on and the fuel that we are running out of most quickly, is the most politically vulnerable of our principal energy sources. More than 90 percent of all transportation in the industrial world depends upon petroleum products. With the end of the petroleum era suddenly in sight, world transport must soon change fundamentally. The problem of fitting transportation into our energy budget is not merely one of designing more efficient systems; we need new systems designed to survive the aftermath of the oil age.

The coming metamorphosis in transportation will involve more than trading in our present technologies for new ones. The contemporary world has been shaped, to a greater extent than is generally realized, by transportation systems based upon cheap oil. Our present patterns of industrial development, of urban organization, of agricultural production, and of recreation are all petroleum products. As oil grows more expensive, eventually becoming too dear to burn, change will be inevitable. The only question is whether such change will be anticipated and brought about smoothly.

A gentle weaning from petroleum will not be possible as long as people tend to view increases in transportation volume as signs of progress. Measured this way, the world has made enormous advances in recent decades. On a finite globe, however, it is possible to go only so far before one begins running in circles. "The prime purpose of passenger transportation is not to increase the amount of physical movement," according to urbanologist Lewis Mumford, "but to increase the possibilities for human association, cooperation, personal intercourse, and choice." Futurist Hazel Henderson goes one step further, suggesting that the volume of transportation may serve as an index of dysfunctional social organization. In a well-planned society, people should not have to travel long distances between their homes and their workplaces, favorite shops, and recreational centers.

More important than determining the relative merits of buses and subways, or of diesel motors and Stirling engines, is the need to structure societies in ways that reduce the need for travel. Some insight into this potential can be gained by comparing two industrial nations with different transportation mentalities. West Germans log only *half* as many passenger miles per capita as Americans. Since German fuel consumption per passenger mile is also about half that in the United States, West Germany uses only 27 percent as much fuel per capita on passenger transport as does the United States. Moreover, although American freight transport uses less fuel per ton-mile than the German system, the United States ships five times as many ton-miles of freight per capita as does West Germany. The sheer volume moved in America overwhelms the advantage in efficiency.[1]

Such volume is desirable only if it is unavoidable. Societies could be fashioned to minimize transportation needs. But for the last several decades, cheap transportation has been substituted for thoughtful design.

Modern cities, for example, have been built on the evanescent foundation of cheap oil. Roads, garages, parking lots, service stations, and car dealerships occupy more than half the land in most large metropolitan areas. The flow of food, clothing, medicines, and other goods is lubricated by oil. Not accidentally, the greatest growth of the world's cities occurred in lockstep with the expansion of the world oil trade. Enormous investments undergird these cities that, unfortunately, grew out of the conditions of yesterday and are maladapted for tomorrow.

The price of oil, corrected for inflation, declined between the late 1940s and the early 1970s. The price of urban real estate shot up like a balloon without a string during the same period, owing to migration from rural areas and to the growth of central business districts. As cheap oil was substituted for expensive land, people found it economical to live in distant suburbs where land was cheaper and to commute long distances each day. Those who could not afford to live in the suburbs were crowded into peripheral slums in numbers greater than the water, sewage, education, and transportation systems could handle. The average distance traveled from home to work stretched. Urban expansion occurred in concentric rings that could not be efficiently served by public transportation systems; swollen thoroughfares formed thromboses of individual vehicles that threatened the survival of downtown areas.

Large, sprawling cities are not, by most lights, attractive places in which to reside. Pollution, congestion, traffic perils, and frayed nerves are synonymous with urban life the world over. As public transit deteriorates, those who cannot afford cars are left more stranded than ever. Too much of what should be free time is spent trying to get somewhere. Although cities can offer attractive economies of scale and extraordinary opportunities for human interaction, these features characterize cities much smaller than most of the world's principal metropolitan areas. The energy costs and the myriad other problems of life in the big city often grow faster than the population (and the tax base), creating a vicious vortex of urban disintegration.

One alternative to continued unplanned urban expansion is the conscious development of "new towns." A recent study for the U.S. Council on Environmental Quality indicated that planned communities require only half the gasoline of typical "sprawl" communities. New towns afford a fresh start, a chance to profit from the mistakes of the past and to experiment with new types of human settlement.[2]

In the new Swedish city of Jarvafaltet, industrial and other employment opportunities are confined to a linear area seven kilometers long and one kilometer wide. This lengthy "downtown" is flanked on both sides by parallel strips of housing interspersed with recreation and shopping facilities. Workplaces thus do not intrude into living areas, but each house is relatively close to the narrow employment corridor. Travel inside this corridor is easily accomplished by public transit. Spinal growth is organic, and in linear developments the corridors can be lengthened, easily and incrementally, as the city outgrows its strip. Jarvafaltet has not banned cars, but it has consciously eliminated the advantages of car-owning.[3]

A radically different approach is being taken at Milton Keynes, an English new town. Elaborate computer simulations indicated to the Milton Keynes planners that dispersed, as opposed to centralized, employment would hold many advantages for the townspeople. In a decentralized city, each person could reside in the immediate vicinity of his or her job, and have the necessities and amenities of life clustered close at hand. And at Stevenage, located thirty miles from London, living, working, and shopping areas have been successfully integrated with twenty-five miles of bikeways and a town center reserved strictly for pedestrians. Studies have shown that Stevenage residents travel much less than people in conventional English cities. Three out of four trips in Stevenage cover less than two miles—a handy distance for walking or biking. Only one out of ten trips exceeds five miles.

Transportation requirements can also be minimized by substituting communication for travel. The energy needed to complete a telephone call is a trifling fraction of the energy needed to transport a person by car or jet. A 1975 report by the U.S. Department of Commerce found that 16 percent of urban automobile transportation in the United States could be replaced by the use of existing telecommunications. New techniques, including facsimile transmission, computer communications, and closed-circuit television can often be substituted for transportation when information rather than materials needs to be exchanged.

Some communications visionaries see recent technical advances as the leading edge of a fundamentally different form of social organization. Marshall McLuhan writes of a "global village" and Peter Goldmark advocates a "new rural society"; each of them envisions more decentralized forms of social organization in which information, education, business transactions, and cultural events can all be transmitted to and from the far corners of the earth. Many of the apparent advantages that led to the growth of major metropolitan centers pale in the light of new communications possibilities. Should the theorists prove correct, new social organizations requiring less transportation and offering opportunities for greater utilization of solar energy resources may evolve over the coming decades. The return of large numbers of people to small, rural communities has become an important trend in many industrial countries, and a governmental objective in much of the Third World.

Of course, urban redesign and "global villages" provide few short-term solutions to the problem of limited transportation energy. Even in the long term, regardless of how intelligently we restructure our cities, or how assiduously we substitute communication for travel, substantial transportation requirements will remain. These needs cannot long be met by gasoline-powered internal-combustion engines; remaining petroleum supplies are too meager, and synthetic petroleum substitutes will be too expensive. Nevertheless, the gasoline-powered automobile is not likely to disengage its clutch overnight. Therefore, opportunities for increased energy efficiency in automobiles—improved mileage and increased occupancy—need to be examined, as do future alternatives to the automobile.

The automobile is the basic unit of the modern, industrial transportation system. This chrome-bedecked symbol of affluence is being embraced worldwide as rapidly as rising national incomes permit. Even the People's Republic of China is negotiating to manufacture a German automobile.

Between 1960 and 1970, the world's population increased less than 20 percent, but the number of automobiles increased more than 100 percent. With a quarter billion cars, we now have one car for every sixteen people inhabiting the earth.

That 100 percent growth spurt was not geographically even. Europe and North America accounted for nearly three-fourths—the rest of the world divided the remainder. Nearly as striking as the differences among countries were the differences between urban and rural areas. Bangkok has three-fourths of the cars in Thailand; Nairobi has 60 percent of Kenya's cars; and Teheran has half the cars in Iran. São Paulo, with one car for every six persons, has about the same car-people ratio as New York City.

The car cult has reached its zenith in the United States. Today the United States has more licensed drivers than registered voters, and two cars are delivered for every baby born. Motor vehicle and allied industries account for one out of every six jobs. In one way or another, the automobile absorbs more than one-fifth of the total U.S. energy budget.

Detroit's enthusiasm for big, powerful, full-optioned cars is easy to understand.[4] Car manufacturers do not sell transportation; they sell vehicles: the more expensive the vehicle, the greater their financial return. Price has traditionally been correlated with size, and no effort has been spared to persuade Americans to trade up to larger, more impressive machines. No particular rationale supports this pricing pattern. The principal costs of manufacturing—labor and overhead—are almost the same for all cars, large and small. But a tradition developed of selling large cars at high profits, and until recently much of the public had been confused into equating size with quality.

For the last two decades, the American automobile industry has steadfastly bred behemoths. Consequently, when the Arab oil embargo was announced, Detroit had no new small cars on its drawing boards. General Motors borrowed a mini-car already in production in Europe and South America and rushed it into the 1976 domestic lineup as the Chevette. The thrifty Chevette soon became the modern equivalent of an earlier American fuel-saver, the Tennessee Walking Horse. Bred for an efficient gait, the animal was sold with the slogan: "A Walker goes further, faster, and saves enough oats to get back again."

The automobile has changed little in the past half century, even though the world through which it travels has changed enormously. Compare, for example, that new Chevette with a predecessor. A typical moderately priced 1915 car weighed a ton or less, and had a four-cylinder, four-stroke, water-cooled, front-mounted engine that powered the rear wheels through a drive shaft. It had a manual transmission with three forward gears and reverse. All these were standard features on the 1975 Chevette. With the exception of the automatic transmission, introduced on London buses in 1926, the automobile industry has not come up with a major innovation in the last sixty years. The Chevette's engine is larger, of course, but the 1915 car could exceed all of today's speed limits. Indeed, an automobile race held in 1908 was won by a car averaging 128 miles per hour.

The evolution of the automobile, considered from the viewpoint of energy efficiency, has been almost entirely maladaptive. Cars tend to be oversized, overpowered, and encumbered with a multitude of accessories, most of which consume lots of fuel to help the driver avoid trifling muscular or mental exertions. For example, to avoid occasionally moving their feet and hands a few inches, many drivers pay extra for automatic transmissions that decrease gasoline mileage by 10 percent or more.

The car facilitated modern metropolitan sprawl, but it is not always beloved in the world it helped to make. As the urban environment has gradually changed, hostility to the traditional automobile has mounted. The respected French opinion poll SOFRES found that 62 percent of the French favored banning cars from central cities. More than a hundred European cities have created auto-free downtown shopping areas.

Yet, what would we do without cars? It is hard to imagine Turin without Fiat, Wolfsburg without Volkswagen. Ninety percent of the families of Coventry, England, rely upon the manufacture of cars and car parts for a livelihood. Closely linked to such other industrial giants as the oil and steel industries—with change in one rippling through all —the automobile industry has become one of the strongest conservative forces in modern society.

Dramatic change is in the wind; faltering oil resources guarantee it. Yet to date the automobile industry does not appear to recognize its altered circumstances or to be preparing seriously for the post-petroleum age. Some legislatures, on the other hand, are mandating minimum levels of fuel economy, and a political debate over how far the shift toward increased mileage can be pushed has begun in several countries.

The physicist's conception of the efficient vehicle is one that operates without friction. At a steady speed on a level road, it would consume no energy. Energy used for acceleration would be recovered during braking; energy used for climbing hills would be recovered when descending. In the real world, of course, friction cannot be avoided: engine parts rub one another; tires encounter road resistance; and the chassis must push its way through resisting air. But car manufacturers could approximate the physicist's ideal much more closely than they do—

witness the 377 miles per gallon achieved in the Shell Mileage Marathon for automobiles.

Abandoning automatic transmissions would save one-tenth of automotive fuel use. Switching to radial tires would save another tenth. Since fuel consumption decreases about 2.8 percent for each 100 pounds of weight reduction, reducing the size of the average American vehicle from 3,600 pounds to 2,700 pounds would save one-quarter of the United States' present gasoline use. A further reduction to 1,800 pounds would reduce automobile fuel needs by nearly half. These smaller cars would require smaller engines, which would cut fuel requirements still more.[5]

A number of strategies and devices could be used to curb fuel waste without curbing vehicles entirely. Streamlining automobile bodies would greatly reduce air drag. For trucks, the potential fuel savings from improved aerodynamic design alone have been estimated at from 20 to 30 percent. Slowing down on the highway will produce much the same result, since air resistance increases exponentially when vehicles travel at high speeds. Installing better ignition systems could save much of the 7 percent of all automobile gasoline now wasted while cars idle. Moreover, using new technologies such as flywheels could help us save much of the fuel (30 percent of all that is used in urban driving) that is dissipated in braking. Avoiding rapid acceleration and quick braking would be even better, for calm, steady driving conserves fuel.

Room for similar improvement in automobile options abounds. For example, cars are at present so poorly insulated that they require air conditioners capable of cooling a small house. Insulation should be substantially improved, and absorption air conditioners for automobiles should be designed to run on waste heat from car engines.

A great many improvements, some of them quite imaginative, have been suggested for the internal-combustion engine. Regardless of such first aid, however, this inefficient and inherently polluting engine faces a bleak future. Eventually, it will run out of gas. Before then, it must be replaced with a more efficient engine that does not guzzle a petroleum-based fuel.

One alternative to the internal-combustion engine that has captured intermittent attention since the beginning of the auto age is the Rankine cycle, or "steam," engine. Few external-combustion engines use water these days, and research is proceeding on various other fluids with superior operating characteristics. Now only its relative bulkiness and long warm-up time need to be reduced. Unfortunately, because the steam engine failed to compete effectively many decades ago, none of the major auto manufacturers takes the Rankine cycle seriously today.

The Brayton cycle, or gas turbine, engine can run on almost any combustible liquid. Already widely used on modern airliners, the Brayton cycle could be scaled down for use in personal transport. Existing turbines require expensive precious metal alloys for some key parts, but these may be replaced by ceramics. Research on the gas turbine has been most vigorously carried out by the Chrysler Corporation.

The Stirling engine has improved significantly since it was patented in 1816 by Robert Stirling, a Scottish clergyman. Most recent advances have been tied to the research of the large Dutch company, N. V. Philips, which has outfitted several Swedish buses with Stirling engines. The engine employs an external flame to heat gas in a closed system, which expands to power a piston. In the new improved version, heat is then removed from the gas by a regenerator and stored to reheat the gas during the next cycle. The efficient reuse of heat gives the Stirling engine its fuel efficiency, and the clean external flame produces far fewer emissions than the explosive firing of the internal-combustion engine. The New Concepts Research Office of the Ford Motor Company has become intrigued with this "old concept," and Ford executives hope that their company will produce a commercial Stirling no later than 1985.

Some believe that Ford and its competitors are purposely dragging their feet. A detailed technology assessment made in 1975 by an independent team of senior engineers from the Jet Propulsion Laboratory of the California Institute of Technology urged that a billion-dollar transition from internal-combustion engines to Stirling and gas turbine engines be made rapidly. The conversion of the U.S. automobile fleet alone could save two million barrels of fuel per day by 1985, the study found. If the fuel saved were gasoline, the savings would amount to more than $8 billion a year at today's fuel prices. Either the gas turbine or the Stirling engine is expected to cost only about $200 more than an internal-combustion engine of the same size, and this investment would be rapidly recovered in fuel savings. Moreover, both engines could operate on fuels ranging from peanut oil to perfume, including such possible gasoline substitutes as ethanol, methanol, and hydrogen.

Hydrogen looks particularly attractive as a transportation fuel. It can be obtained by breaking down water using several different renewable energy sources, and its combustion residue is pure water. However, hydrogen is hard to store for use in cars: it is difficult to liquefy, so large volumes of hydrogen gas are needed if much energy is to be stored. Interesting research is now being done on storing hydrogen as hydrides, metallic compounds that release the gas when heated.

Electric cars have champions more powerful than they are, including most of the electrical utility industry. Utilities at present need to make major capital investments that produce power to meet "peak" daytime demand but that remain idle at other hours. Widespread adoption of electric cars would mitigate this problem, as most cars would operate on their batteries during daylight hours and be recharged at night when there is idle capacity. However, cost estimates for electric cars generally neglect the fact that batteries are "consumed," that they wear out, and that their depreciation generally costs more per kilowatt-hour than electricity. A study by the Stanford Research Institute found that batteries used in electric cars cost about ten cents a kilowatt-hour to run, excluding the cost of recharging.

Widespread use of electric vehicles would confine pollution to the power plant and free us from dependence upon petroleum fuels. But, to date, such cars have limited battery storage (hence limited range), perform poorly in cold weather, reach only modest speeds, and accelerate slowly. New generations of batteries, including lithium sulfur and silver zinc batteries, may overcome these difficulties. In 1974, for example, a motorcycle powered by a silver zinc battery set a speed record of 160 miles per hour. But many of these new batteries are prohibitively expensive.

Batteries are particularly attractive for delivery vehicles, because conventional motors waste much of their fuel while idling. American Motors has a contract to provide the U.S. Post Office with 352 electric delivery vehicles, and about 50,000 electric vans are operating in Great Britain. Various hybrids of electric cars with other auxiliary motors have also been proposed.

Flywheel propulsion is yet another way to go. Now found in devices ranging from sewing machines to spacecraft, flywheels smooth out uneven power cycles by providing steady momentum. The principle of the flywheel is simply that a turning wheel with low-friction bearings stores mechanical energy. When energy is put in, the wheel turns faster; after energy is withdrawn, the wheel turns more slowly. The amount of energy stored is a function of the weight of the wheel and the rate of rotation. Since the storage capacity increases exponentially with the rotation speed, a little more spin stores a lot more energy. Flywheels can be used as brakes, storing the energy that would otherwise be lost as a vehicle decelerates, and then feeding power back out when the vehicle gets under way again.

Most American innovation has been associated with the aerospace effort, although a couple of recent projects brought flywheels back to earth. The U.S. Energy Research and Development Administration funded a $200,000 project to apply flywheels to automobiles, and the U.S. Urban Mass Transit Administration invested about $2 million in applying flywheels to subway trains and trolley buses. The Soviet Union has also done extensive flywheel research.

Most cars are as inefficient as a person who sleeps twenty-three hours a day. Automobile commuters pay a sizable fraction of their disposable income for vehicles that may be used one hour or less each day; for the rest of the time each car unproductively occupies as much land area as a standard office. Taxis are one partial solution, and jitneys—cars or small vans that follow a fixed route at frequent intervals—are another. In 1915, about 60,000 jitneys operated in U.S. cities. Streetcars eventually forced them out of business, and taxi companies have successfully lobbied to keep them out. In many other countries, however, jitneys frequently provide a cheap and relatively efficient form of public transit.

A bolder solution to the problems of under-utilized vehicles is the Witkar, designed to ameliorate the traffic problems of Amsterdam. This seven-hundred-year-old Dutch city predates not only the automobile but also the horse and buggy era. The resulting traffic problems defy the imagination. Of the 35,000 automobiles that enter Amsterdam every day, only 4 percent are moving at any time; the remainder are parked along the streets and in lots. The Witkar is designed to keep vehicles in circulation through joint ownership and use. The electric-powered vehicles can be picked up at any recharging station by anyone who has paid his $10 lifetime membership fee. The rider is charged about four cents a minute for the vehicle until he returns it to another recharging station. The Witkar, as at present designed, has a top speed of 18 mph and a range of about two miles. It seats two comfortably with additional room for packages or a small child. Currently, only thirty-five Witkars and five recharging stations are operating, but three thousand people have already paid their lifetime membership fees. The Witkar is only an experiment, and its future is not assured even in Amsterdam. Nonetheless, it is an example of the surge of bold innovation desperately needed by cities everywhere.

The best all-round alternative to the automobile for short trips is probably the bicycle. Cycling is several times more efficient than walking: a cyclist traveling at 10 mph uses only 100 Btu's per passenger-mile, while a pedestrian walking at 2.5 mph uses 500 Btu's per mile. The cyclist obtains the energy equivalent of 1,000 passenger-miles per gallon —noticeably better than most sub-compacts—and consumes food, not petroleum. If, following Ivan Illich's suggestion, we attribute to the automobile not only the time spent behind the steering wheel, but also all the time spent earning money to purchase, maintain, fuel, and insure a typical car, and compare that aggregate figure to an equivalent number for a bicycle, the bicycle emerges as considerably faster for all urban trips.[6]

Perhaps the most bicycle-conscious country is the Netherlands, with 11 million bicycles for 13 million people. Five million Dutch students, workers, and others bicycle daily, rain or shine. Although bikes are much more popular in rural areas than in cities (where cyclists fear the dangers of pollution and heavy traffic), former Dutch Transport Minister William Drees believes that "the bicycle could return as the main means of urban transportation in six to eight years." What would be required, in Drees' view, are overpasses and special lanes to protect cyclists from motorists. In Rotterdam, bicycles already account for more than a quarter of all trips made using any form of transportation.

The advantages of bicycles speak for themselves. A bicycle requires only one-thirtieth the space of a large car—a crucial factor in congested urban settings. It provides an opportunity for exercise, thus helping the rider relax nervous tensions, shed excess poundage, and maintain good health. Bicycles consume no non-renewable resources and they produce no pollution.

The bicycle is an elegantly simple device. Developed in its modern, chain-driven form less than a century ago, the technology has been transferred around the world with almost unique success. Inexpensive and easily maintained, it is equally at home in elite suburbs and pleasant pathways. It can carry formidable loads, especially in such three-wheeled variations as the Trishaw and the Vendor. Bicycles can generally exceed most urban speed limits.

Unfortunately, the contemporary city was built around automobiles, and bikers compete at their peril. An estimated one million bicyclists require medical attention in the United States each year, many for accidents involving automobiles. Eleven hundred U.S. bicyclists were killed in 1973. Millions of other cyclists temper the beneficial health effects of cycling with the risks of accelerated respiration of air rich in lead, hydrocarbons, carbon monoxide, and asbestos particulates. Meanwhile, the population at large enjoys the cleaner air made possible by those who cycle instead of drive.

In the rain, cycling can be miserable. First, the bicycle has no roof. Second, modern science has been unable to produce a bicycle brake that is reliable in wet weather. These factors discourage bicycle use where it would otherwise appear attractive. As a leading Bombay transportation planner told me, "Our people are too poor to buy vehicles they cannot use for one-third of the year. There are few bicycles on the street today; when the monsoon comes, there will be none."

If the bicycle is again to play an important role in the transportation field, its drawbacks must be lessened. In most new towns, exclusive bicycle lanes are incorporated in the over-all design. In some existing cities, whole streets have been dedicated to bicycles. The weather problem is less easily solved. Besides rain gear, proposed answers include canopied bike paths and lightweight two-person pedicabs with roofs.

A bicycle-motorcycle hybrid, the moped, is popular in Europe and in some Asian countries. Over 32 million have been sold worldwide. The moped is basically a bicycle rigged up with a 1- or 2-horsepower engine, capable of powering the vehicle at 20 to 30 mph. It costs under $500, and runs up to 200 miles on a gallon. European accident statistics suggest that mopeds are less safe than bicycles, but less perilous than motorcycles.

More than half of all automobile trips are less than five miles long, even though automobiles perform at their worst in the short stint because cold engines are relatively inefficient. Some studies suggest that fuel mileage on four-mile trips is less than two-thirds that obtained when the engine is warm. For such short trips, bicycles and mopeds would hold a substantial advantage, if only our cities were designed so that they could be safely and comfortably used.

Although the energy efficiency of individual vehicles is undeniably important, vehicle occupancy may deserve even more attention. In the United States, intercity cars contain an average of 2.4 persons, intracity cars hold an average of 1.4 persons at a time, and rush hour commuter vehicles carry an average of only 1.2 passengers each. Fifty-six percent of all American commuters currently drive to work alone, while 26 percent share cars with others; 14 percent use public transportation; and 4 percent walk, bicycle, or use other means. Automobile passenger lists have persistently shrunk. Former U.S. Environmental Administrator William Ruckelshaus once jokingly predicted that "at existing rates of automobile passenger decline, by 1980 one out of three operating vehicles will not have a driver."

Meaningful statements about comparative modes of transportation cannot be made without first making some assumptions about vehicle occupancy, known among transportation planners as the "load factor." Almost overnight, conservation-minded, automobile-dependent countries could double or triple the average load factor of automobiles. Commuting lends itself particularly well to such car pooling.[7]

If other modes of transport replace the automobile, considerable savings can accrue. A switch can be made to twelve-passenger commuter vans or mini-buses that operate as car pools or group taxis but that carry more passengers per mile than either. Such vehicles are widely and successfully used in Peru. Scores of U.S. companies provide commuter vans for their employees, and these are proving economical and popular. The companies find that it is cheaper to buy a van than to maintain parking spaces for a dozen individual vehicles; the commuters find that the operating expenses they assume are much lower than the expenses car ownership entails.

Demand-responsive transportation systems are also being tried in many cities. Dial-a-Ride, Dial-a-Bus, Telebus, and others are all similar in operation. Riders telephone a control center, giving their location and destination. They are then grouped with other riders with similar origins and destinations. A radio-dispatched vehicle picks them all up and takes them from doorstep to doorstep more cheaply and efficiently than could a taxi carrying only one passenger. Such systems are being used in forty American and Canadian cities.

. The U.S. Department of Transportation encourages the development of "people movers," or personal urban rapid transit systems. People movers consist of many small automatically controlled vehicles that carry passengers along a fixed track. Now used widely to carry passengers between airline terminals and sightseers around zoos, the people mover is a sort of horizontal elevator. The passenger climbs aboard the vehicle, punches a button to indicate his destination, and a central computer sends the car on its way. For short runs along fixed routes, personal rapid transit systems are probably inferior to rail transit lines that can haul ten times as many passengers during peak hours, and can adjust to non-peak demand by shortening the train and running less frequently. A demonstration unit built at Morgantown, West Virginia, has been plagued with operating difficulties and expensive cost overruns.

In Europe, the trolley is a traditional and long popular form of transportation. In the United States, home of the world's most famous streetcar, the trolley has just about disappeared. In 1932, the General Motors Corporation formed a subsidiary for the purpose of purchasing streetcar companies, tearing up their tracks, dismantling their power lines, and replacing streetcars with GM buses that do their polluting downtown. Over the subsequent two decades, GM, with help from the oil and tire industries, "motorized" electric rail-trolley bus lines in forty-five cities, including New York, Philadelphia, Baltimore, St. Louis, and Los Angeles.[8] Today, the trolley, reincarnated as the light rail vehicle, is staging a comeback. With a much lower carrying capacity than railroads and subways, trolleys are best suited to cities of one million or less.

Vienna's superb trolley system is serving as a model for Milan and other Italian cities. Mexico City's 250 streetcars and 550 trolley buses carry 250 million passengers a year. Boston and San Francisco recently placed orders for modern trolleys, and Dayton, Ohio, is planning a comprehensive new trolley system.

The comparative merits of different public transit systems are a matter of continuing controversy. A number of glittering generalizations can be made, but, even when true, they can be wildly misleading. For example, buses are about twice as energy-efficient per seat-mile as automobiles, but not, as one might suspect, because of their weight. A bus with the same luggage capacity per passenger as an automobile weighs *more* per passenger seat than do small automobiles; a commuter bus with no luggage compartment weighs only slightly less. The principal advantages of a bus are its high-pressure tires and its small diesel engine. But automobiles could, of course, be equipped with harder tires and smaller engines, as many now are.

Rail systems might be expected to be much more efficient than either buses or cars. The rolling resistance of a steel wheel on a steel track is many times less than rubber on asphalt, and the aerodynamic drag on a train is less than for cars. However, these theoretical advantages are generally lost in practice. Rail systems tend to achieve much higher speeds than buses and cars, and to lose the energy spent on acceleration in braking. Speed also increases aerodynamic drag (which is proportional to the square of the velocity, so doubling the speed quadruples the resistance). Drag is also much greater in subway tunnels than on the surface. Moreover, the heating, cooling, and lighting requirements for rail systems are substantial. About half of all energy used in the San Francisco BART system is for heating, air conditioning, and station lighting. Indeed, BART consumes about as much energy per seat-mile as a typical automobile.

Public systems might be expected to have higher load factors than automobiles. During peak periods, they do. The Tokyo subway system pays uniformed men to shove rush hour commuters into jammed cars so that the doors can close; in winter, jackets and overcoats worsen the crunch but provide the small comfort of a cushion. In Bombay, the load factors of rush hour commuter trains have been estimated at about 500 percent of designed capacity. However, calculating load factors is tricky. Automobiles leave one area with all their passengers and arrive at a destination with their passengers aboard. Public vehicles start out empty and gradually fill up during the course of their route. The "average" load factor may be only 50 percent. After they arrive downtown during the morning rush and discharge their passengers, they often have to return to the outlying area nearly empty. Public systems must operate during non-peak hours, and average load factors are much lower then. Although data are somewhat sketchy, several U.S. studies indicate that public transit load factors average between 18 and 25 percent of capacity—roughly comparable to automobiles.

To be sure, things other than energy must be considered. Riding on BART is more comfortable than riding in most automobiles; it is several times faster than commuting by car; and it doesn't pollute the downtown area. On the other hand, this pleasant, high-speed transportation option may encourage people to live farther away from their workplaces than they otherwise would. BART serves only 4 percent of Bay Area commuters and does so at a substantial subsidy. Equally sobering, the development it triggered in downtown San Francisco drove up real-estate prices, forcing more urban residents, particularly poor people, to move to the outskirts and become commuters.

If designed to utilize their technical potential to save energy, if operated at high load factors, and if coupled with a successful campaign to eliminate the one car/one driver syndrome, public transit systems can enhance urban life. But they can do so only within the context of a comprehensive transportation plan that has as one of its highest priorities the minimization of over-all transport volume. In the past, too many partisans have mistakenly viewed mass transit as a simple solution to all urban transportation problems. Any "solution" must be as much social and political as technical. The federal government of the United States expects to spend about $12 billion on mass transit and about $20 billion on highways between 1975 and 1980; other countries will also invest enormous sums on transportation. These investments represent great commitments of scarce capital, and they will shape the world's cities for years to come. Thus, such spending must be guided by a comprehensive vision of how we want those cities to look.

Intercity travel can also be made more economical and more efficient. Throughout much of the world, in fact, a rather energy-efficient rail transportation system is now being built. However, in the United States during the 1960s, railroad passenger traffic declined by half, automobile mileage increased by half, and air passenger traffic tripled.

Clocked only in terms of miles per hour, trains cannot keep up with airplanes. But for shorter trips of up to four hundred miles, railroads can actually save travel time. Picking up and delivering passengers right downtown, trains eliminate the need for costly, time-consuming taxi rides to and from airports. In addition, trains never circle a city waiting for inclement weather to pass, nor are they ever diverted to a landing strip a hundred miles away from their intended destination by a snow-storm.

Where railroad systems have been made a national priority, they have proven effective. Japan's bullet trains race back and forth between Tokyo and Osaka sixty times each day, averaging 101 mph and topping 125 mph for some portions of the journey. The Japanese high-speed rail network is being expanded—it already stretches to Hiroshima—and is being engineered to accommodate speeds of up to 155 mph.

Western Europe is served by forty-four plush Trans-Europe Express trains connecting 185 cities in ten countries. Three-fourths of all German track is welded rail, and France has 3,000 miles of welded track, to allow the use of high-speed trains. French trains average over 90 mph between Paris and Bordeaux, and a line scheduled to be opened in 1982 between Paris and Lyon will average more than 130 mph. Italy is completing a rail link between Rome and Florence, on which 100-mph trains will run.

A problem for some railroad or airplane travelers with rural destinations is a transportation tie-up at their journey's end. In rural destinations, rental cars and taxis are expensive. One solution to this problem is to place their cars aboard auto-passenger trains. At the destination, the driver's personal car is unloaded as he disembarks from the coach. France has fifty-seven such auto trains.

For trips longer than several hundred miles, railroads are at a disadvantage for people who value their time. But for vacations, travel by train can be relaxing.

The least energy-efficient mode of transport between cities is by supersonic jet or by private "executive" jet. The Anglo-French Concorde burns somewhat over 5,500 gallons of fuel an hour, while carrying fewer than a hundred passengers. Unable to fly long distances, impractical for short flights, banned from flying at supersonic speeds over most land areas, and perhaps contributing to the depletion of the ozone layer, these intermediate-range "prestige" planes are the ultimate example of transportation technology run amuck. Executive jets, because of their small load factors, require about as much fuel per passenger-mile as the Concorde. Large conventional jets, scheduled at reasonable intervals to ensure high load factors, are a far preferable form of high-speed travel.

Freight Transport

In addition to moving people, the transportation system also hauls goods. The energy efficiency of freight transport varies widely among modes—hovercraft and helicopters rank lowest, while supertankers, barges, and pipelines are several times more energy-efficient than trucks.[9] In the United States, trucks haul less than one-fifth of all freight but consume about one-half of all fuel expended on freight transport. Pipelines, waterways, and railroads carry more than 80 percent of all freight but consume less fuel combined than do trucks alone.

Of course, waterways and railroad tracks do not extend to most neighborhood department stores, or even to many regional warehouses. But an ideal freight-hauling system would assign each task to the mode that performs it most efficiently. Packing cargo in containers allows such intermodal transfer to be accomplished rather easily, and in other cases the "piggybacking" of truck trailers on railroad flatcars is an efficient alternative. Unfortunately, in much of the industrialized world, trucks have been replacing trains even for long-distance hauls, for which they are poorly suited. Between 1950 and 1970, the percentage of total ton-miles hauled by U.S. railroads declined from 47 to 35 percent, while the equivalent figure for trucks rose from 13 to 19 percent. The fastest-growing sector (although still a comparatively minor one) has been air freight. While air transport has no equal when speed is essential, its increasing use for shipments having no time constraints is inexcusably wasteful.[10]

Creative freight transport experts have suggested resurrecting old technologies. For certain transport tasks, airships (dirigibles) appear to have significant energy advantages. Because they expend no energy keeping themselves and their cargo aloft, airships require a small fraction of the energy needed by airplanes. A study by the Southern California Aviation Council indicated that airships could haul freight for 2,000 miles at 100 mph at a cost of 4.4 cents per ton-mile: cheaper than air or truck, but more expensive than railroads. Ghana is currently experimenting with a German-built zeppelin devised to haul freight to inaccessible locations. Airships can deliver directly to any destination, hovering overhead as their cargo is unloaded. However, world supplies of helium are limited, and the *Hindenburg* disaster dramatized the danger of replacing helium with large volumes of combustible hydrogen.[11]

For ocean freight, modern sailing vessels might be quite competitive with conventional boats. During the last two decades, international seaborne trade has increased about sixfold, and shipping now consumes more than 100 million tons of petroleum each year. Future volumes will almost certainly shrink as petroleum reserves dwindle and as nations or groups of nations necessarily become relatively more self-sufficient in both energy and food. Nonetheless, oceangoing vessels will be the most energy-efficient means of conducting essential international trade. Hence, we must find a replacement for petroleum as a source of power for seagoing vessels.

Most of the writing on petroleum substitutes has focused on nuclear power. However, a nuclear-powered merchant marine would have technical and political difficulties. The Japanese have had waves of recurrent difficulties with their first nuclear-powered ship, the *Mutsu*, leaving many other countries leery of such vessels. With anti-nuclear sentiment seemingly on the rise around the world, nuclear ships would also run the risk of being banned from certain ports and waterways. Finally, the nature and costs of nuclear reactors make them attractive possibilities only for mammoth vessels, and they thus have little potential for use in small and intermediate ships.

Marine history may well repeat itself. Coal was once the marine fuel of choice, and it could again become significant. But in the long run coal supplies will also be exhausted. Fuels derived from biomass, such as methanol, may offer some promise. But the most fascinating suggestion is doubtless the return of the sailing vessel. The wind carried wayfarers across the oceans for millennia before steamships displaced sailboats in the early twentieth century. Although the most rapid development of sailing vessels occurred in the nineteenth century, under competitive threat from steamships, they were eventually doomed by a lack of reliability. But now, incorporating the knowledge of commercial sailing acquired during the last century, recent developments from recreational sailing, and advances in the fields of meteorology, aerodynamics, and control engineering, a modern commercial sailing vessel (with auxiliary power for calm periods and for maneuvering in harbors) could compete well against oil-powered ships. Studies at the University of Hamburg, the University of Newcastle upon Tyne, and the University of Michigan found that large modern sailing ships, driven by vertical aerofoils and taking full advantage of modern weather- and wave-forecasting capabilities, could transport freight speedily and reliably while consuming only 5 to 10 percent as much fuel as a conventional vessel.[12]

Transportation is an exceptionally difficult field in which to implement new ideas.[13] A free marketplace often leads individuals to gratify their immediate self-interest, at group expense, thereby creating a situation in which all suffer. Governmental subsidies, incentives, and regulations—each with its supportive private vested interest—so thoroughly riddle most transportation networks that bureaucratic reform requires great political muscle. The problems of different modes are generally approached in a piecemeal fashion, and comprehensive transportation plans thus fail to take shape.

Often transportation innovations fall short of their objectives, and sometimes they fall flat on their faces. For example, the rationale most frequently given for the construction of mass transit systems is to reduce the volume of automobile traffic. Yet experience indicates that most mass transit riders are not former automobile drivers but former bus passengers, walkers, automobile passengers, and homebodies. Two years after the Mexico City subway opened, it was overloaded. Yet street congestion was not improved, because most transfers came from buses. In addition, new businesses located along the subway line greatly increased over-all travel demand along that corridor.

The central goal of an efficient urban transportation system should be to eliminate, or at least control, the one-person-to-a-car system. A variety of cures have been suggested. An increase in the price of fuel will probably be slow to make itself felt. Prices will rise automatically as petroleum supplies decline, or they can be raised through taxation. In Sweden, a 60-cents-per-gallon gasoline tax has been rather successful at reducing one-person vehicles to a minimum.

Gas rationing accomplishes the same result with a somewhat heavier hand. If a central authority reduces the amount of gasoline available by one-third, car drivers will consume one-third less. Rationing does cost money to administer, unlike taxes which raise revenue. But neither gasoline taxes nor gasoline rationing discriminate against a particular time of day, type of vehicle, or number of passengers (although both measures could reasonably be expected to lead to smaller cars and higher load factors).

Congestion pricing has been tried successfully in some situations. When all traffic must pass through a specific corridor, such as a bridge or a tunnel, it is possible to collect a toll and to vary the charge with the time of day and/or the number of passengers. In San Francisco and New York, variable bridge tolls have proven viable. In Singapore, a different kind of congestion pricing is used. A limited number of rather expensive stickers, which allow automobiles free access to otherwise restricted sections of the city, are sold. Any vehicle in those areas not displaying a sticker is fined. This effectively places a ceiling on vehicle use in congested sections of town.

Designated lanes limited to use by car pools and buses have also effectively encouraged higher load factors. They motivate drivers grinding their teeth in traffic jams to switch to one of the multiple-passenger vehicles whizzing past in exclusive adjoining lanes.

Parking controls are still another means of restraining automobile use. Some businesses have banned the use of their parking lots to employees not using car pools, and many have arranged their lots so that single-driver vehicles must park far afield. Some cities have imposed stiff parking lot taxes—San Francisco's is 10 percent and Pittsburgh's is 20 percent—in an effort to reduce the number of commuter automobiles entering the downtown area. Parking taxes also allow cities to recover some of the indirect costs of commuter automobiles that otherwise fall on urban taxpayers.

A final resort is to ban automobiles altogether from certain streets or sections. Nagoya, Japan, a city of two million, uses this approach (in combination with preferential treatment for public transportation vehicles) with great success. Many smaller cities and towns around the world, including many of the medieval towns of Italy, have enacted limited bans on automobiles. Travel in the car-free areas is limited to pedestrians, bicyclists, and public transportation vehicles, all of which flow smoothly and rapidly instead of inching their way through snarled traffic.

While discouraging use of the automobile in which the driver is the sole passenger, the transportation system must provide alternatives for those who have abandoned their cars. Transit systems should be attractive, reasonably priced, and intermeshed in terms of both physical hookups and timetables. Controls over land use must also be vigorously exercised in order to make living near work a practical possibility.

An integrated approach to transportation is needed—one that eliminates unnecessary travel while using a multitude of incentives and penalties to make necessary travel efficient. This will not be easy to accomplish as we simultaneously begin to wean ourselves from oil. The process must be begun, and quickly, unless the world is to grind to a standstill at the end of the petroleum era.

7. Btu's and Buildings:
Energy and Shelter

PEOPLE ARE fragile creatures, dependent more upon wits than physical endowment for survival. Although the surface temperature of our planet fluctuates more than 150 degrees Centigrade, we cannot endure more than a five-degree variation in blood temperature. Birds migrate and bears hibernate to cope with chill winds. People insist on business as usual, and build shelters against the storm, the sun, the rain, the wind, and the cold.

From earliest recorded history, we have sought refuge from the elements in shelters, over which we have gradually learned to exercise a high degree of control. In times past, this control was a response to nature; it encompassed the careful use of appropriate building materials and proper orientation to natural features, including the sun, prevailing winds, and local terrain. But in recent decades we have begun to use massive amounts of energy to control interior space. Between 1950 and 1970, for example, the energy requirements per square foot in new office buildings in New York City more than doubled. All buildings—north and south, mountain and desert—now tend to resemble one another; moreover, they are nearly identical on all four sides, seeming to ignore entirely the existence of the sun. Only in the entrails, in the relative sizes of furnaces and air conditioners, is the external world taken into account at all.

Shelters warm and illuminate our winter nights, cleanse and chill polluted summer air, and shield us from spring rains. But such necessary protections are increasingly purchased at an unnecessarily extravagant energy cost. The American Institute of Architects has estimated that:

If the U.S. adopted at high-priority national program emphasizing energy efficient buildings, we could by 1990 be saving the equivalent of more than 12.5 million barrels of petroleum per day. . . . We are now investing vast quantities of increasingly scarce capital resources in strategies which have less potential, less certainty and longer-delayed payoffs than the proposed alternative strategy emphasizing a national program for energy efficient buildings.

Elsewhere, the AIA explicitly notes that "the decision is not whether to modify functional demand or behavior or level of comfort; rather it is whether to invest capital to waste energy or to utilize that same capital to conserve energy."[1]

Energy is seldom a criterion in the selection of building materials, though use of less energy-intensive materials need not entail sacrifice of either strength or durability. Stainless steel, for example, can often substitute for aluminum. Although somewhat more steel than aluminum is generally required for most building purposes, the energy cost of refining a pound of steel is only one-fifth that for a pound of aluminum; Richard Stein, chairman of the New York Board of Architecture, calculates that the 2 million kilowatt-hours of electricity needed to refine enough aluminum to sheathe a building the size of Sears Tower in Chicago could be cut by two-thirds with a switch to stainless steel. Glass, a terrible insulator, may be the least desirable building material from an energy viewpoint. Even double-paned glass, if not exposed to sunlight, can lose ten times as much heat as a well-insulated wall; losses through a single glazed window could be twenty times as great. Additionally, the windows in most glass buildings cannot be opened, so the energy cost of introducing fresh air is hefty.[2]

The priority concern in designing energy-efficient buildings is to minimize the transfer of heat between the structure and its environment. Comprehensive efforts to reduce heat transfer have resulted in fuel savings of up to 80 percent in some buildings. Attempts to control heat loss and heat entry must take into account the three ways that heat moves in and out of buildings: through conduction, convection, and radiation. Each requires a different heat-management technique.

Conduction refers to heat transfer in solids. If one end of an iron pipe is placed in a fire, the other end soon grows hot. Heat is carried along the pipe's length by conduction. Since heat losses and heat gains are directly proportional to the amount of surface exposed, the ratio of external walls to internal space should usually be kept at a minimum. And, since heat transfers by conduction are inversely proportional to the thickness of the conducting material, doubling the thickness of a wall cuts the potential amount of heat it can transfer by half.

Materials that conduct heat poorly can be used as insulation. Because the transfer of heat between solids is directly proportional to the temperature differences of the two surfaces, the larger the temperature gap between the interior and outside air, the more insulation should be installed. Most buildings in the temperate zone are under-insulated, and too many—especially among the dwellings of poor people—have no insulation at all. Investments in insulating such structures will—dollar for dollar—generally save several times as much energy as investments in new mines and power plants will produce.

Convection occurs when air circulates between a building and its exterior environment. Pressure differences—caused, for example, by wind, furnaces, or ventilation equipment—force air (and thus heat) in and out of the structure. The amount of heat that air can carry off is astonishing: a quarter-inch crack along a three-foot attic door can cost more than 20 gallons of fuel oil during a moderate winter. In almost all U.S. houses, more than one-half of the building's volume of air escapes each hour; often the leakage rate is two or three times higher. Air seepage around doors and windows can be reduced by weather stripping, caulking, storm doors and windows, and double-glazed glass. In commercial buildings, installing vestibules or revolving doors will often provide warmer welcomes and reduce both air circulation and fuel bills. Many public buildings ventilate several times more air than is necessary to maintain internal freshness. New air must be either heated or cooled and is often also "scrubbed" to remove pollutants. All incoming air is super-chilled and then partially reheated as necessary in many large buildings —a technique known as "terminal reheating" that uses far more energy than necessary. Since the object is to heat and cool buildings, not cities, the flow of air through such ill-begotten buildings should be kept to a minimum.

Radiant sunlight is the boon or the bane to the "climate" of most buildings, depending upon how it is used. It can be captured by solar collectors and used to regulate the building's temperature as desired, or it can be admitted carelessly, especially through windows, shackling fuel-consuming temperature-control equipment with an extra burden.

Unwanted solar radiation can be screened by keeping window areas small, using awnings or shutters, planting shade trees, using tinted or reflective glass, or employing light colors on roofs and walls. Adjustable exterior window shields called Rolladen that are a hybrid of shutters, venetian blinds, and awnings shelter the windows of many European buildings.

Energy conservation measures are often thought of as add-on expenses that can be amortized over many years through reduced fuel bills. However, some practices save money right from the start. For example, the Toledo Edison Building uses double-paned glass with a chromium coating to reflect heat; it cost $122,000 more to install than standard quarter-inch plate glass. However, the energy-conserving glass enabled engineers to reduce the building's heating plant by 53 percent, the cooling system by 65 percent, and the distribution systems by 68 percent, for a gross initial savings of $123,000, and a net savings of $1,000. Even more attractive in economic terms, annual operating costs are $40,000 lower than they would have been had conventional glass been used.

The two hundred energy-saving houses constructed in Arkansas under a grant from the U.S. Department of Housing and Urban Development cost no more to construct than two hundred houses built using standard construction techniques. Annual heating and cooling bills for these dwellings, however, were only one-fourth the size of those for comparable conventional houses in the neighborhood.[3]

A $30,000 investment in one five-story building at Ohio State University reduced the structure's subsequent natural gas consumption by 78 percent and its electricity use by 43 percent, for an annual savings of $60,000. The repayment period was six months.

Sometimes the basic choices that determine energy use levels involve no cash outlays at all. Choosing to live in buildings that share walls and thus have lower energy requirements than detached structures is just one example. A recent report compared the energy budget of a typical suburban "sprawl" community with the energy needs of a planned community having mostly town houses and apartments. With reduced spatial needs and fewer exposed walls, the planned community required only half as much natural gas and two-thirds as much electricity as did the sprawl community.[4]

Buildings, like transportation systems, have "load factor" energy efficiencies. Five passengers in a single automobile can each travel to and from work far more economically than a solitary driver can. Similarly, a fully occupied building is more energy-efficient than one that goes mostly unused. Yet more than half the office space built in New York in recent years remains empty. Employees who work overtime in huge buildings with centralized services can impose particularly dire drains; a worker who insists upon air-conditioning his hermetically sealed office in the World Trade Center must cool thirty-one floors on that face of the building. In like manner, many families heat and cool large areas of unused residential space.

Regardless of how thriftily a building is designed and operated, it will, of course, require some energy. A number of rather sophisticated systems, district heating operations among them, have been designed to provide this energy as efficiently as possible. Widely used in Europe, district heating schemes allow the centralized use of fuels such as coal and garbage that could be difficult to burn cleanly in individual urban structures, and transfer the heat efficiently to where it is used. About two hundred European cities warm buildings with the heat from incinerated trash; many more use coal. Geothermal district heating is used in parts of Iceland, New Zealand, and France.

District heating provides a wise use for the large quantities of waste heat cast off by electrical power plants and by some industries. Its basic design is that of a closed loop that takes hot water from a power station to a consumer and returns it to the main plant for reheating. The water coursed through the loop might reach 212 degrees Fahrenheit at the station, register 206 degrees when it arrives at the consumer's heat exchangers, leave the consumer at 132 degrees, and arrive at the central plant at 130 degrees, where it is reheated to 212 degrees. Such a system provides two or more times as much building heat per unit of fuel consumed as do setups that generate electricity or synthetic gas to run electrical resistance heaters or gas furnaces.

Total energy systems, which generate on-site electricity and use "waste" heat to both heat and cool buildings, can be even more efficient than district heating. Among the advantages they confer is a degree of independence from centralized power grids. Moreover, because the electrical generation takes place near where the electricity will be used, transmission costs and losses can be slashed. During the large-scale electric blackout of the northeastern United States in 1965, many newspapers ran a photograph of one cluster of well-lit buildings amidst the darkness of Queens. The buildings in the island of light were served by the Rochdale Village Cooperative's 20,000-kilowatt total energy system.

Modular integrated utility systems (MIUSs) are a refinement of the kind of total energy system that weathered the blackout in New York. They attempt to integrate all utilities—electricity, heating and cooling, waste disposal, and water—into one efficient package. Whereas most total energy systems are custom designed, MIUS systems will consist of interchangeable and mass-produced modules. A Jersey City, New Jersey, MIUS system powered by five 600-kilowatt generators provides electricity, heat, air conditioning, and hot water for six large apartment buildings, a school, and a 50,000-square-foot commercial building.

Clearly the ideal source of energy for building operations is direct sunlight. Since the overwhelming bulk of the average building's energy requirement—70 percent or more—is for low-grade heat, rather elementary solar equipment will suffice. Literally hundreds of different techniques can be utilized to harness diffuse solar energy to meet a building's needs.

Solar heating systems for buildings can be either "active" or "passive." In active systems, fans and pumps move air or liquid from a collector first to a storage area and then to where it is needed. Passive systems store energy right where sunlight impinges on the building's structural mass; such systems are designed to shield the structure from unwanted summer heat while capturing and retaining the sun's warmth during the colder months. Passive solar buildings act as "thermal flywheels," smoothing the effects of outside temperature fluctuations between day and night—a principle as old as the ancient thick-walled structures of Mohenjo-Daro in the Indus Valley and the adobe Indian pueblos in the American Southwest. Although more money and attention has been lavished upon active systems, many of the world's most successful solar buildings employ simple, inexpensive passive designs.[5]

In the latitudes that girdle the earth between 35 degrees north and 35 degrees south, roofs of buildings can be built to serve as passive solar storage devices. For this region, American designer Harold Hay has built a "sky-therm" house, the flat roof of which is covered by large polyethylene bags filled with water. By adroitly manipulating slabs of insulation over the roof during the day or night, Hay can heat the house in the winter and cool it in summer. A. K. N. Reddy and K. K. Prasad at the Indian Institute of Science in Bangalore have suggested a similar but less expensive design for poor countries; their model uses rooftop ponds of water.

In latitudes above 35 degrees either north or south, a flat roof can catch less and less of the low winter sun. Vertical walls and steep roofs are more effective solar collectors in these regions than are flat roofs. In France, Felix Trombe and Jacques Michel have built several solar houses, each with a glass wall facing south and a thick concrete wall located a short distance inside the glass. Openings near the top and bottom of the concrete walls create a natural circulation pattern as hot air rises and moves into the living areas while cool air flows through the bottom opening into the solar-heated space between the glass and the concrete. During the summer, when additional heat is unwanted, the top air passages are closed and the rising air is channeled outside. This same approach has been successfully employed by Doug Kelbaugh in his passive solar house in Princeton, New Jersey.

Steve Baer, one of the cleverest American solar inventors, has incorporated a unique passive solar system that stores sunlight in barrels in his New Mexico house.[6] On the indoor side of a glass wall, Baer has stacked 91 metal barrels filled with 4,800 gallons of water. The drums store considerable heat, and an interesting pattern of sunlight enters the room around their edges. Outside the vertical slab of glass, Baer has placed another wall, made of lightweight insulation sandwiched between sheets of aluminum. This outer wall is hinged at the bottom and can be easily raised or lowered. When erect, say on a winter night or summer day, the outer wall can keep heat either in or out of the building. When lowered to allow the sun to strike the barrels, the inner aluminum sheet acts as a reflector, causing sunlight that would otherwise strike the ground to rebound against the drums.

In addition to such passive approaches, hundreds of active solar heating systems have been built, using a variety of collectors and storage systems. Each technology stresses certain features—good performance, rugged durability, attractive appearance, or low cost—each of which is often achieved at the sacrifice of others. The U.S. effort has been by far the most expensive and ambitious, though important work has been done in the Soviet Union, Great Britain, Australia, Japan, Denmark, Egypt, and Israel.

Flat-plate solar collectors suffice for normal heating purposes, and can either be made by the builder from available materials or mass-produced rather cheaply. For very high temperatures, such as those needed to power some absorption air conditioners, costlier collectors that use selective surfaces or focusing devices to track the sun across the sky are needed. After heat has been collected and then transported to storage reservoirs, most active solar heating systems use conventional technologies (water radiators or forced-air ducts) to deliver it to the living areas as needed.

Storing heat for a couple of days is not difficult; heated water or gravel will do the job if a large insulated storage bin is used. Eutectic salts, substances that absorb prodigious amounts of heat when they melt and then release it when they re-solidify, can reduce the minimum storage volume needed by a factor of six. The most serious problems plaguing the storage of heat in phase-changing eutectic salts have been overcome, according to Dr. Maria Telkes, a leading American expert in solar thermal storage.[7]

In the 1940s, the Japanese built an energy storage system that worked on an annual cycle. During cold months, heat was pumped from a large container of water; by the end of the winter, a huge block of ice had formed, into which excess building heat was dumped during the summer. The Japanese concept was recently revived by Harry Fischer of the Oak Ridge National Laboratory in Tennessee. Fischer found that when combined with a solar collector, a radiator, and an efficient heat pump, such an annual storage system can perform admirably over a wide range of climates. Fischer's prototype worked so well that several private companies decided to develop the concept further.[8]

Many simple solar technologies can be used to cool buildings. Simple ceiling vents may suffice to expel hot air, at the same time drawing cooler air up from a basement or well. In dry climates, evaporative cooling can be used to chill the air. In more humid areas, solar absorption air conditioners may be needed. The logical successors to contemporary cooling units, solar air conditioners are currently being developed in Japan and the United States. While early solar air conditioners required heat at about 120 degrees Centigrade for optimum performance, a

Japanese company has developed a unit that operates satisfactorily at 75 degrees Centigrade—a temperature any commercial solar collector can easily muster. Fortuitously, solar air conditioners reach peak cooling capacity when the sun burns brightest, which is when they are most needed. Consequently, solar air conditioners could reduce peak demands on many electrical power grids. As cost-effective solar air conditioners reach the market, the over-all economics of solar systems will improve because the collectors will begin providing a year-round benefit.[9]

It is harder in temperate than in tropical regions to provide with solar technologies 100 percent of the heat buildings need. It is generally cheaper at present to get supplementary heat during long cloudy periods from conventional fuels, wind power, biogas, or wood. However, when solar equipment is mass-produced, prices should plummet, while fossil fuel prices can only climb. Moreover, major improvements in the design of collectors, thermal storage systems, and heat-transfer mechanisms are being made. Indeed, the day is dawning when heating and cooling self-sufficiency will be an economical option for most new buildings.

Solar heating systems are most attractive when considered in terms of "lifetime costs"; the initial investment *plus* the lifetime operating costs of solar systems often total less than the combined purchase and operating costs of conventional heating systems. For example, recent U.S. studies have shown solar heating to be more economical than electrical heating except in competition with cheap hydropower.[10]

Investments in solar technologies can be mortgaged at a steady cost over the years, while the fuel costs of alternative systems will rise at least as fast as general inflation. In fact, the initial cost alone of solar heating systems often amounts to less than the initial cost of electrical resistance heating, if the cost of the building's share of a new power plant and the electrical distribution system is included. However, the cost of a solar heating system must be borne entirely by the homeowner, while a utility builds the power plant and strings the power lines. The utility borrows money at a lower interest rate than the homeowner can obtain, and it averages the cost of electricity from the expensive new plant with power from cheap plants built decades earlier so that true marginal costs are never compared.[11]

Solar-heated buildings are now commercially viable. However, large-scale changes in the housing industry are not accomplished easily—

witness the 30,000 autonomous building code jurisdictions in the United States. The building industry is localized, with even the giant construction firms producing fewer than one-half of one percent of all units. Profit margins are small, and salability has traditionally reflected the builder's ability to keep purchase prices low. Nonetheless, a respected market research organization, Frost & Sullivan, predicts that 2.5 million U.S. residences will be solar-heated and cooled by 1985, and the American Institute of Architects has urged an even more ambitious solar development program.[12]

Solar heating becomes even more attractive when it is crossbred with other compatible technologies. Its happy marriage to absorption air conditioners and heat pumps has already been mentioned. Greenhouses too can be splendid solar collectors, producing much more heat than they need in even the dead of winter, if they are tightly constructed, well insulated, and fitted with substantial thermal storage capacity. Whereas many old-style attached greenhouses placed demands on the heating system of the main house, inexpensive solar greenhouses can actually furnish heat to the living area while they extend the growing season for home-grown vegetables. A program to build greenhouses for low-income families in northern New Mexico out of local materials, low-cost fiberglass, and polyethylene has already proven successful.

Solar photovoltaic cells, which generate electricity directly from sunlight, are still too expensive for most homeowners. Prices have dropped precipitously in recent years, however, and many specialists predict that these non-polluting, decentralized units will compete economically with centralized power plants within a few years. Photovoltaic cells can convert only about one-fifth of the energy in sunlight into electricity, but the remaining four-fifths need not be wasted; a photovoltaic cell can be used in tandem with an active solar heating system, which can collect the remaining energy as heat. Highly thermodynamically efficient, a combined system also consumes no fuels, produces no pollution, and relies upon no large utility grid.

Although most of the energy used in buildings goes for heating and cooling, almost 30 percent serves other purposes. In commercial buildings, lighting usually claims most of the remainder, with a variety of machines accounting for the rest. In residences, food storage and preparation command a significant fraction. Pilot lights on gas ovens, which ought sensibly to be replaced by electric igniters, account for more than 40 percent of the fuel such ovens consume. Stoves can make things hot for refrigerators and should be placed at some distance from their antagonists. Similarly, placing the refrigerator against an exterior wall (but not in direct sunlight) allows waste heat to be easily vented outside in summer or to be retained inside during the winter.

The electrical use in most visible need of improvement is lighting, which in the United States consumes about a fourth of all electricity sold. Some controversy exists over just what level of illumination is necessary and desirable, but some enlightened thinkers suggest that prevailing standards are almost always higher than those necessary for optimal performance. Corroborating this view is the fact that lighting levels in the home, where personal choice can be exercised, are far lower than the average levels in commercial buildings and schools.

Compounding the waste of radiant light is the widespread tendency to light unused space. Lights are seldom focused solely on work spaces; instead, large rooms—or even whole floors of buildings—are unnecessarily lit up. Joseph Swidler, former chairman of the New York Public Service Commission, once noted that the corridor outside his office had "more than enough light for fine needlework, miniature painting, or engraving counterfeit money, although it was used only for walking from office to office."

Since fluorescent bulbs deliver three to four times as much light per unit of electricity as incandescent bulbs do, enormous amounts of energy could be saved by switching from filament to fluorescent bulbs. But even fluorescent bulbs convert only about one-fifth of the electricity they use into light. A NATO-sponsored scientific committee on energy conservation reported that there is "no fundamental theoretical reason why a 100 percent conversion efficiency" could not be attained.

Light bulbs shed the electricity that they do not convert into light directly as heat. Reducing the lighting level in buildings and switching to more efficient bulbs thus reduces the size of the needed cooling system, and lowers the initial cost of installing light fixtures and wiring. For an air-conditioned building, every two watts of unnecessary lighting necessitates the use of one additional watt for cooling. Over half the air-conditioning load in many office buildings is needed to combat the heat generated by lights.

The appearance of sustainable new fuel-conserving technologies on the market is, of course, of little interest to people who cannot afford even the old technologies. Xerxes Desai, former general manager of the Indian "new town" of New Bombay, notes, "You can't save much energy in Bombay's buildings. We don't require heating, and the fraction of our population that can afford any kind of air conditioning—solar or otherwise—is microscopic."

It is difficult to exaggerate the differences in the options available to the rich and poor. Today, the fraction of Americans who have air conditioning use more electricity for that one purpose than 800 million Chinese use for everything. The United States lavishes more electricity on lighting than is generated by the continents of Africa, Asia, and South America combined.

Although much of the Third World is located in climates that require no heating, other parts—especially mountainous regions—depend heavily upon the burning of firewood and dung for warmth. People in all climates, of course, need fuel to cook and to heat water. Elementary solar technologies based on devices easily made and easily maintained can, along with biogas and wind and water power, provide even the poor with the beginnings of energy self-sufficiency, especially if they live in small towns or rural areas. An efficient stove for cooking and heating, topped by a simple pressure cooker, could double or triple the benefits many poor families squeeze from existing fuel supplies. Even among the poorest, efficiency must be a prime concern, and they must be given the chance to be efficient.

Energy is destined to play an increasingly visible role in the shelters of all people everywhere. While some are utilizing the latest advances in photovoltaic technology, others will be reasserting the ancient wisdom of planting shade trees and windbreaks, of harnessing prevailing winds for ventilation, and of relying on thick ceilings and walls to even out daily extremes in temperature. The most successful in all cultures will be those who realize that we have reached the end of an era, and who design shelters to work with nature instead of defying it.

8. Energy and Economic Growth

ALTHOUGH SOME scientists may exaggerate or underestimate planetary limits, most of them understand that such physical limits exist. Economists, on the other hand, tend to reject the concept out of hand. This difference of opinion is understandable in light of the way the current generation of economists has invested its intellectual capital. Most economic analyses blur the important distinctions between economic growth and physical growth. Yet economists, few of whom have paid serious attention to the social and physical constraints on energy growth, make virtually all energy demand forecasts.

Economic growth and physical growth are not synonymous. The physical growth with which chemists and biologists concern themselves is measured in physical terms: grams, meters, watts, joules, and so on. But the growth with which economists are concerned is the increase in the *value* of commodities and services. Thus, economic growth is not limited by the finite nature of the physical world. Growth can be accomplished by changing designated values, or even by redefining terms. By taking into national income accounts the goods produced by do-it-yourselfers, or by valuing the services performed by housewives, or by charging admission to a park that was previously free, or by selling pollution control equipment that merely remedies a cost that was previously borne unwittingly by society, one can argue that economic "growth" has occurred.

Total growth in the production of goods and services has tended to outpace growth in the use of fuels and materials, partly because economic growth in many developed nations is occurring most rapidly in the economic sectors—services—that are dominant in a "post-industrial" state.[1] Services, which include leisure activities, education, and health care, tend to require far less energy and far fewer materials than the production of goods. Another more subtle part of the explanation is that much economic growth is attributable to the assignation of monetary value to the "quality" of the factors of production. Because of improvements in this qualitative dimension (e.g., better-trained workers, more productive technologies, innovations in management, etc.), industry can now obtain more units of output per unit of input.

To the extent that economic growth reflects only growth in value, it can continue almost indefinitely. But to the extent that economic growth is rooted in a physical dimension, it will be subject to physical limits. Economic analyses based on historic relationships between fuel consumption and economic growth will prove to be increasingly in error as these limits begin to assert themselves. Economic growth can continue indefinitely only if it can be successfully divorced from energy growth.

Industrial Energy Use and Abuse

Industrial forecasters often mistake rearview mirrors for crystal balls. Their tomorrows look remarkably like yesterday—only bigger. Hindsight seems to carry more weight than foresight in planned as well as in capitalist economies, in many small firms as well as in behemoths. Not surprisingly, then, those seeking to transform energy policy generally view industry as an enemy to progress. Applied to those industries whose business is to whet and then to fill the nation's growing appetite for fuel, this view is accurate. Yet other industries may wind up playing a positive role in the coming energy transition.

More than most other parts of society, industry carefully analyzes the long-term implications of its expenditures, often in thirty- to fifty-year time frames. It watches lifetime as well as initial costs hawkishly, and dares to make sizable investments when those investments reduce its operating costs.[2] Finally, industry values security highly, and, consequently, values an energy source whose reliability and price can be predicted for the foreseeable future. Thus, self-interest alone may well prompt industry to embrace renewable energy resources, and energy-conserving measures.

Coal will doubtless play a vital role for industry in the transition period. Because coal is relatively bountiful and has less value than petroleum as a chemical feedstock, the substitution of coal for fuel oil makes sense in many industrial processes. In addition, new combustion technologies, including fluidized beds, will reduce the environmental consequences of burning coal. However, wide-scale coal use entails grave inherent problems, and this fuel should be gradually phased out as renewable energy sources become available.

Shifting industry to dependence upon energy supplies derived from renewable sources will require significant adjustments. But these changes will be much less painful than those made necessary by major nuclear electrification programs or by growing dependence upon synthetic fuels made from coal or shale. Energy sources will have to be deftly matched with appropriate uses; production will become more labor-intensive; plants will be more decentralized than at present; and most production materials will be reused or recycled. Such changes may easily be viewed as desirable in themselves, apart from their importance in the shift to the use of renewable sources.

Industry has led the way in world energy conservation efforts over the last few years. When the efficacy of full-scale sustainable energy systems has been proven to their satisfaction, industrial decision-makers may welcome the new technologies much more rapidly than is commonly expected.

Energy Efficiency in Industry—Enlightened Self-Interest

Nicholas Georgescu-Roegen, an unconventional economist who has assessed the role of energy in economic production,[3] notes that "there is a difference between what goes into the economic process and what comes out of it." Since the process cannot create matter or energy, the "difference" is that matter has been rearranged to serve human needs or wants. Such work requires the use of energy, but the energy actually *required* is ordinarily a small fraction of that now spent. Lamentably, the thermodynamic efficiency of U.S. industry is only about 25 percent.

Things could be different. The West German steel and petroleum industries use only two-thirds as much energy per ton of product as do their American counterparts; the German paper industry uses only 57 percent as much per ton as the U.S. paper industry. Yet the scope for improved energy efficiency even in German industry is enormous.[4]

Comparisons between countries and between different facilities in the same country make it clear that reducing industrial fuel consumption need not reduce economic output. Consumption cutbacks require only the increased use of fuel-efficient industrial machinery and the improved operation and maintenance of this machinery. Cutbacks may also lead to the substitution of labor and capital for fuel and to a shift in the mixture of goods and services produced. For the past fifty years in the industrialized world, the amount of fuel consumed per dollar's worth of goods and services produced has fallen—despite declining real energy prices. With rising energy prices a near certainty for the foreseeable future, this trend could accelerate dramatically.

Energy conservation has traditionally been among industry's lowest priorities.[5] Fuel has been so inexpensive that extravagant fuel use has gone unquestioned; moreover, energy prices (adjusted for inflation) fell steadily for decades, and popular mythology held that future sources would eventually be "too cheap to meter." Industrial energy efficiency has nonetheless improved over the years, mostly through rather unimaginative advances. Energy conservation has simply not attracted large numbers of the most talented researchers. Charles Berg, former chief engineer of the U.S. Federal Power Commission, has noted that ". . . the application of greater insulation on water-cooled furnace skid rails to save fuel is unlikely to stimulate greatly the curiosity of the young student physicist or engineer, or his professor." René Dubos, the microbiologist and philosopher, goes so far as to argue that the industrialized world's current "overuse of energy tends to interfere with the adaptive and creative mechanisms of response that are inherent in human nature and external nature."

Conserving industrial energy used to mean just eliminating embarrassing waste. For example, when it had infrared photographs taken of a facility to detect heat leaks, the Dow Chemical Company discovered that a sidewalk heating system used to clear pathways of snow had been left on in summer. The company "conserved" energy by flipping a switch that had been left on by accident. Other companies accomplished major savings by repairing broken windows and closing huge, two-story factory doors.

The biggest opportunities for fuel savings, however, require more sophistication.[6] Devices such as recuperators, regenerators, heat wheels, and heat pipes, for example, help conserve the heat generated in industrial plants, heat that would otherwise be used once and discharged or removed directly with the flue gases without having been used at all.

Particularly impressive gains can be made in the primary metals industries.[7] Energy savings of over 50 percent can be made in the steel industry if older plants are gradually replaced by more efficient facilities. For example, continuous casting holds a large energy advantage over ingot pouring, and major differences exist in the efficiencies of different types of blast furnaces. In addition, hot coke is at present often quenched with water—a method that wastes energy while producing enormous amounts of air and water pollution. In plants in Europe and the Soviet Union, coke is cooled with a recycled inert gas, and much of its heat is recaptured to perform useful work.

The manufacture of aluminum is so energy-intensive that the industry has generally situated its major installations near sources of large amounts of cheap electricity. Technical advances in the traditional Hall aluminum refining process can reduce energy requirements by more than a fifth; Alcoa is now building a major facility using a new chloride process that is expected to reduce energy needs by almost one-third. The Aluminum Research Corporation of New Orleans is developing a new chemical process that should use even less energy than the Alcoa process does.

The paper and cement industries also waste energy. The most efficient paper-manufacturing technologies require 50 percent less fuel than other commonly used methods need. If, in addition to embracing more efficient conventional technologies, industry were to use wood wastes as fuel, Swedish-style, some paper factories' demand for fossil fuels could be slashed by an astonishing 75 percent. In cement manufacturing in the United States, an average of 1.2 million Btu's is used to decompose enough limestone to produce a barrel of cement. In some European plants, where waste heat from cement kilns is recaptured to preheat the limestone feedstock, only 550,000 Btu's are needed per barrel.

An important part of increasing the energy efficiency of industry will be matching energy sources of different qualities to appropriate uses. The lower-grade heat that remains after high-grade energy is used should be recaptured and used to perform other work. This process of using energy at each of the thermodynamic stages of decreasing usefulness through which it passes is sometimes termed "cascading."

At present, electricity fulfills much of industry's energy demand. In the United States, electricity constitutes about one-third of all industrial energy, and most of this electricity is purchased from large centralized power plants. The average efficiency of American power plants is below 30 percent; fully 70 percent of the energy originally contained in the fuel they use is discharged into the environment as low-grade heat. But factories have many needs for low-grade heat, needs they now meet by burning high-grade fuels. If electrical generation took place inside factories instead of at remote power plants, the waste heat could be efficiently cascaded through multiple uses.

For an industry producing high-pressure steam, the amount of additional fuel needed to produce electricity is only half that needed to generate electricity in a central power station. A study performed for the National Science Foundation recommended that the United States install at least 50,000 megawatts of industrial co-generation capacity by 1985. The study pointed out that such investments produce a minimum annual return of 20 percent, and require only half as much capital per unit of energy produced as do investments in new centralized power plants.[8]

Industrial energy conservation is not always cheap to implement. The capital required for major retooling in industry can, on the contrary, sometimes be substantial. Because society does not have an endless supply of capital, major investments of one type necessarily foreclose other options. Consequently, competition exists between the financial requirements of new energy facilities and the capital needed for improvements in industrial energy efficiency. For example, the original United States proposal for Project Independence would have required $1 trillion by 1985, four-fifths of which would have been earmarked for new, rather than replacement, energy facilities. This commitment would have claimed a full two-thirds of all new net capital investment during that period—money that would otherwise have been spent on other industries, transportation, housing, and so forth. Major investments must be made in all these sectors if they are to convert to more energy-efficient processes. The pool of available capital is limited, and large-scale investments in new energy facilities can be made only by using money that could more fruitfully be invested in increased efficiency. If, as Gregory Bateson contended, capital is the "stored flexibility" necessary for any structural transformation, society would greatly narrow its industrial options by investing too heavily in new energy facilities.

Energy versus Jobs

Major investments in new energy facilities, it is often said, will contribute to full employment. The Executive Council of the American Federation of Labor has called for sustained energy growth to promote "high employment, a dynamic economy, and a satisfying way of life." However, new energy facilities are among the least labor-intensive investments a society can make. Capital diverted from nuclear reactors, coal gasification facilities, and petroleum refineries will produce more employment if invested in almost any other enterprise. Insulating homes provides far more jobs per investment dollar than building petroleum refineries to produce heating oil does, and the money the homeowner saves every year on fuel bills will provide additional employment when spent on food, clothing, recreation, or health care.[9]

By utilizing techniques that substitute skilled labor for energy, great fuel savings can be built into the construction industry. Richard Stein, chairman of the New York Board for Architecture, has criticized the current "trend toward construction techniques which substitute masses of material for more careful design and construction." Stein calculates that the electricity used in manufacturing *unnecessary* cement alone amounts to about 20 billion kilowatt-hours a year—over a fourth of the electricity produced annually in India. With the employment of more and better labor to mix and place cement correctly, use of this material could be halved.[10]

The factors of industrial production—labor, energy, materials, and capital—are, within limits, interchangeable, and can thus be arranged in various combinations. The relative productivity of any factor varies with its cost. The argument over whether labor is productive because it is high priced or high priced because it is productive is academic. If labor is expensive compared to capital and energy, machinery and megawatts will be substituted for muscle wherever possible. The history of industrial development has been, in large measure, a history of the substitution of energy, capital, and materials for labor.

As a general rule, the more energy-intensive a product is, the less labor-intensive it will be. As another general rule, services (other than transportation) require more labor and less energy than do physical commodities. Fuel and electricity are, of course, the most energy-intensive commodities on the market. They provide more energy per dollar —and fewer jobs per dollar—than anything else for sale. Thus, to the extent that industry conserves fuel and spends its fuel budget on *anything* else, employment will rise.

"Over the years we have substituted energy-powered capital equipment for people because the work can be done more efficiently and at a lower cost," notes John Winger, Executive Vice-President for Energy Affairs at the Chase Manhattan Bank. Winger then concludes that "we can't turn back; we couldn't afford to." But his conclusion is not self-evident, and the end of the development tunnel—of which his vision is a part—could be dark.

"The Coming Age of Automatic Factors," an article printed in *Technology Review* in early 1975, predicted that "complete manufacturing systems governed by central computers will be demonstrated by 1985." The magazine quotes a leading automation company executive as stating that if present trends continue, only 2 percent of the U.S. labor force will be engaged in manufacturing in the year 2000. With 2 percent of its labor force engaged in manufacturing, the United States would obviously have a great many jobless citizens. Obviously, such projections are absurd, and a prime rationale for an aggressive energy conservation program might be to avoid just such mass unemployment.

Substituting labor for energy would save workers from more than unemployment. As workers have been displaced by machines, a growing economy has in the past been able to provide jobs for most of them. But these jobs, though "productive," tend to lack a qualitative dimension that is important to human dignity. E. F. Schumacher argues that "the type of work which modern technology is most successful in reducing or even eliminating is skillful productive work of human hands, in touch with real materials of one kind or another." Schumacher believes that "modern technology has deprived man of the kind of work he enjoys most . . . and given him plenty of work of a fragmented kind, most of which he does not enjoy at all." Saving energy should not be used as an excuse to resurrect dreary, unrewarding forms of manual labor, like ditchdigging, that are best left to prideless machines. But where human skill, intelligence, or craftsmanship have been replaced by automation, they should again be given reign.

Recycling

If we continue to expand the use of raw materials at present rates, the extraction and processing of minerals and other natural resources would exert ever-increasing pressure upon our energy supplies. In the past, high-grade deposits could be exploited using relatively little energy, but now we are being forced to use lower-grade reserves. Copper, for example, is mined today from ore containing only two-tenths of a percent of the metal, which means that 500 tons of rock must be processed to obtain one ton of copper. The extraction and processing of raw materials, Harvey Brooks estimates, now account for about two-thirds of all U.S. industrial energy use, or about 25 percent of *all* U.S. energy use.

At present, resources are commonly used once and then discarded. In the wealthier countries, these one-way streams have swollen into veritable floods. The American trash heap grows annually by more than 11 million tons of iron and steel, 800,000 tons of aluminum, 400,000 tons of other metals, 13 million tons of glass, and 60 million tons of paper; some 17 billion cans, 38 billion bottles and jars, 7.6 million discarded television sets, and 7 million junked cars and trucks contribute to the totals.

We thus have the option of turning to our garbage dumps for an increasing amount of raw material.[11] The advantages of doing so are manifest. The energy required to produce a ton of steel from urban waste—including separation, transportation, and processing—is only 14 percent of that needed to produce a ton of steel from raw ore. For copper, the figure is about 9 percent; for aluminum, only 5 percent as much energy is needed to recycle the metal as to refine the ore initially.

Even greater savings can generally be realized by repairs and reuse than by recycling.

The recycling of iron, copper, and aluminum in the United States at levels that are now economically practical would save the energy equivalent of 3.3 billion gallons of gasoline each year. Complete recycling would save roughly twice as much. Recycling all steel cans would save the United States as much energy as eight 500-megawatt power plants produce; recycling all paper could, in principle, save energy equal to the annual production of sixteen 500-megawatt plants. If all glass containers were reused six times, the need for nine 500-megawatt power plants would be eliminated.[12]

Using materials again and again reduces environmental wear and tear in many ways. Recycling just one ton of steel, for example, has far-reaching effects. The 200 pounds of air pollutants and 102 pounds of water pollutants associated with refining 2,000 pounds of steel are never released. In addition, the 2.7 tons of mining wastes associated with each ton are never generated and the 6,700 gallons of water needed to refine each ton are never sullied.[13]

Some countries have begun to take advantage of the promise inherent in recycling technology. Leningrad recycles 580,000 tons of garbage each year, producing metal, chemicals, and compost. The Russians plan to expand the facility sixfold by 1985. Large deposits must be paid on glass containers in the Soviet Union; and bottles and jars are reused several times. In Denmark, 80,000 tons of oil and chemical wastes are processed annually at a huge, centralized waste treatment plant. More than 45 percent of paper production in Britain and West Germany now entails use of recycled fibers.

The greatest energy savings occur when unneeded products are taken out of production. For example, a large fraction of all urban trash in industrialized countries consists of packaging that served no useful function before being discarded. Eliminating unnecessary bags and boxes makes far more sense than merely recycling their tattered remains. A cabinet-level report released by the French Minister of Commerce in July, 1975, notes, "It is preferable to incorporate energy and raw materials in an object that lasts a long time rather than manufacture a dozen things to be thrown away almost immediately." The report calls for high taxes on goods with short life spans, including all packaging, and would require manufacturers to supply spare parts for their products.

Decentralization—A Social Frontier

Industries can easily obtain all the energy they need from renewable sources. But direct solar energy, wind and water power, and biological energy sources are more diffuse than conventional fuels. Thus, the high costs of collecting renewable energy and of transporting it to a central location argue for the construction of many small facilities instead of outsized complexes.

It is often said of competitive modern manufacturing that size is of the essence. This misapprehension is doubtless rooted in the fact that large corporations control much of the world's economy. Company size should not, however, be confused with plant size: large firms are almost always clusters of small facilities. As Barry Stein points out, "The very same plant or retail store in a community, depending on whether it is owned by a local entrepreneur or an international conglomerate, shifts in classification from 'small' business to 'large.' " From an energy standpoint, who owns a facility matters less than how big it is.

A surprising fraction of existing manufacturing facilities are relatively small. In the United States, for example, although just 3 percent of all corporations own one-sixth of all plants and employ about three-fourths of all workers, the number of employees in each of these plants averages only 203. If a few assembly-line industries (such as the automobile and electrical equipment manufacturing concerns) are excluded, plant employment among these large multi-unit companies averages about 100. To say that either giant plants or economies of scale don't exist is preposterous. But giant plants are not the norm in most industries, and economies of scale can generally be enjoyed in plants of modest size.[14]

While the question of ownership has little to do with the transition to renewable energy sources, other social advantages do attend decentralized ownership and control. Small firms tend to diversify both wealth and social power; they also seldom wield disproportionate influence over governments. Small firms often provide more room for innovation and for genuine worker participation in decisions, and they tend to be a more integral part of their surrounding communities than their larger counter-parts. Smaller firms also tend to have fewer strikes, better worker-safety records, and less sabotage than large firms. Perhaps partly as a result of all these trends, small firms also tend to generate higher net returns.

The process of industrial decentralization is already well under way, spurred mostly by the desire of people everywhere to escape the pollu-tion and the myriad social ills that blight congested urban areas. An effort toward decentralization in Japan in the early 1970s, prompted by former Prime Minister Tanaka's best-selling book, *Restructuring the Japanese Archipelago*, would have created dozens of new industrial towns of 250,000 people each at a projected cost of $1 trillion. The program stalled when it became public knowledge that the Prime Minis-ter stood to prosper personally from resulting shifts in land value. Nonetheless, decentralizing industry remains one of Japan's top domes-tic priorities. Similar decentralizing trends are visible in France, where the government offered industry a wide range of incentives to locate outside Paris, and in the move of American companies away from the Northeast into the "sun belt." Although such moves are not now being made for the purpose of tapping renewable energy resources, they will make the coming energy transition much easier.

Decentralization is not an economic cure-all. China's abortive expe-rience in the Great Leap Forward should inspire other nations to look before they decentralize certain heavy industries, such as steel. But the steel industry's size requirements make it a special case rather than a test case. In a future powered by renewable sources, energy may be less easily transported than capital, technology, or skilled labor. Industry will con-sequently relocate toward those parts of each country—and, indeed, those portions of the world—where renewable energy sources are in greatest abundance.

III

Safe Sustainable Sources

9. Turning toward the Sun

A BOUT ONE-FIFTH of all energy used around the world now comes from solar resources: wind power, water power, biomass, and direct sunlight. By the year 2000, such renewable energy sources could provide 40 percent of the global energy budget; by 2025, humanity could obtain 75 percent of its energy from solar resources. Such a transition would not be cheap or easy, but its benefits would far outweigh the costs and difficulties. The proposed timetable would require an unprecedented worldwide commitment of resources and talent, but the consequences of failure are similarly unprecedented. Every essential feature of the proposed solar transition has already proven technically viable; if the fifty-year timetable is not met, the roadblocks will have been political —not technical.[1]

Our ancestors captured the sun's energy indirectly by gathering wild vegetation. Their harvest became more reliable with the revolutionary shift to planned cultivation and the domestication of animals. As civilization developed, reliance upon the sun grew increasingly circuitous. Slaves and draft animals provided a roundabout means of harnessing large quantities of photosynthetic energy. Breezes and currents—both solar-powered phenomena—drove mills and invited overseas travel.

In earlier eras, people were intensely aware of the sun as a force in their lives. They constructed buildings to take advantage of prevailing winds and of the angles at which the sun's rays hit the earth. They built industries near streams to make power generation and transport easier. Their lives revolved around the agricultural seasons. In the fourteenth century, coal began to contribute an increasing fraction of Europe's energy budget—a trend that accelerated greatly in the eighteenth and nineteenth centuries. During the past seventy-five years, oil and natural gas became the principal energy sources in the industrialized world. In the fossil fuel era, the sun has been largely ignored. No nation includes the sun in its official energy budget, even though all other energy sources would be reduced to comparative insignificance if it were. We think we heat our homes with fossil fuels, forgetting that without the sun those homes would be −240 degrees Centigrade when we turned on our furnaces. We think we get our light from electricity, forgetting that without the sun the skies would be permanently black.[2]

About 1.5 quadrillion megawatt-hours of solar energy arrive at the earth's outer atmosphere each year. This amount is 28,000 times greater than all the commercial energy used by humankind. Roughly 35 percent of this energy is reflected back into space; another 18 percent is absorbed by the atmosphere and drives the winds; and about 47 percent reaches the earth. No country uses as much energy as is contained in the sunlight that strikes just its buildings. Indeed, the sunshine that falls each year on U.S. roads alone contains twice as much energy as does the fossil fuel used annually by the entire world. The wind power available at prime sites could produce several times more electricity than is currently generated from all sources. Only a fraction of the world's hydropower capacity has been tapped. As much energy could be obtained from biomass each year as fossil fuels currently provide.

How easily and cheaply these vast energy sources can be harvested is disputed. Opinions naturally rest heavily upon the questions asked and the assumptions made. How much distance can separate an energy facility and its potential users? Will people and industries migrate to take advantage of new energy sources? Should only huge, utility-scale sites be considered or should individual and community-sized sites be counted as well? What limits will environmental, political, and aesthetic factors impose?

Past efforts to tap the solar flow have been thwarted by unreasonable economic biases. The environmental costs of conventional fuels, for example, have until recently been largely ignored. If reclamation were required of strip mining companies, if power plants were required to stifle their noxious fumes, if oil tankers were prohibited from fouling the oceans with their toxic discharges, if nuclear advocates were forced to find a safe way to dispose of long-lived radioactive wastes, conventional power sources would cost more and solar equipment would be more economically competitive. As such costs have been increasingly "internalized," conventional sources have grown more expensive and solar alternatives have consequently become more credible.[3]

Moreover, fuel prices long reflected only the costs of discovery, extraction, refining, and delivery; they failed to include the value of the fuel itself. Over the years, improvements in exploitation techniques drove fuel prices relentlessly downward, but these low prices were chimerical. Although, for example, U.S. oil prices (corrected for inflation) fell 37 percent in the twenty-five years between 1948 and 1972, the nation was living off its energy capital during this period—not its interest. The world has only a limited stock of fuel, and it was only a matter of time before that fuel began to run out.[4]

Unlike finite fuels, sunlight is a flow and not a stock. Once a gallon of oil is burned, it is gone forever; but the sun will cast its rays earthward a billion years from now, whether sunshine is harnessed today for human needs or not. Technical improvements in the use of sunlight could lower prices permanently; similar technical improvements in the use of finite fuels can only hasten their exhaustion.

The current world economy was built upon the assumption that its limited resources could be expanded indefinitely. Instead of OPEC-style severance royalties when oil was removed from the earth, depletion allowances were granted to those who exploited it. Instead of a reasonable "scarcity rent" for fuels, the needs of future generations were discounted to near zero. Now that the world's remaining supply of easily obtainable high-grade fuel is mostly in the hands of single-resource nations with legitimate worries about their long-range futures, prices have increased fivefold in five years. As a consequence, solar energy is rapidly shaking off the false economic constraints that previously hindered its commercial development. In 1976, the United States produced one million square feet of solar collectors; in 1977, the figure is expected to triple.[5]

Since sunlight is ubiquitous and can be used in decentralized facilities, many proposed solar options would dispense with the expensive transportation and distribution networks that encumber conventional energy sources.[6] The savings thus obtained can be substantial; transmission and distribution today account for about 70 percent of the cost of providing electricity to the average U.S. residence.[7] In addition, line losses during electrical transmission may amount to several percent of all the energy produced, and the unsightly transmission tendrils that link centralized energy sources to their users are vulnerable to both natural disasters and human sabotage.

Probably the most important element in a successful solar strategy is the thermodynamic matching of appropriate energy sources with compatible uses. The quality of energy sought from the sun and the costs of collecting, converting, and storing that energy usually correlate directly: the higher the desired quality, the higher the cost. Sources and uses must therefore be carefully matched, so that expensive, high-quality energy is not wasted on jobs that do not require it.[8]

The energy currently employed for various tasks is often of far higher quality than necessary. For example, roughly 34 percent of end-use energy in the United States is employed as heat at temperatures under 100 degrees Centigrade; much of this energy heats buildings and provides hot water. Another 24 percent is for heat at temperatures of 100 degrees Centigrade or higher, much of it for industrial processes. Thirty percent of end-use energy is employed to power the transportation system; 8 percent is used as electricity and 3 percent as miscellaneous mechanical work. In Canada, a somewhat higher percentage is used for low-grade heat and somewhat less is used for transportation. Although both countries are highly industrialized, highly mobile, and have high energy use–GNP ratios, most of the energy budgets of both could easily and economically be met using existing solar technologies.[9]

Cheap, unsophisticated collectors can easily provide temperatures up to 100 degrees Centigrade. Selective surfaces—thin, space-age coatings that absorb much sunlight but re-radiate negligible heat—greatly increase the temperatures that collectors can attain. Because air conducts and convects heat, high-temperature collectors are often sealed vacuums. Focusing collectors, which use lenses or mirrors to focus sunlight into a small target area, can obtain still higher temperatures. The French solar furnace at Odeillo, for instance, can reach temperatures of about 3000 degrees Centigrade.

Solar thermal-electric plants appear economically sound, especially when operated just to meet daytime peak demands or when crossbred with existing plants that use other fuels for nighttime power production.

Ocean thermal facilities may be a source of base-load electricity in some coastal areas. Decentralized photovoltaic cells will be the most attractive source of solar electricity if the cost reductions commonly projected materialize.

Wind power can be harnessed directly to generate electricity. But because electricity is difficult to store, some wind turbines might best be used to pump water into reservoirs or to compress air. The air and water can then be released as needed to generate electricity or to perform mechanical work. Energy from intermittent sources like wind machines can also be stored as high-temperature heat or in chemical fuels, flywheels, or electrical batteries.

Biological energy sources, which include both organic wastes and fuel crops, could by themselves yield much of the world's current energy needs. Such sources can provide liquid and gaseous fuels as well as direct heat and electricity. Particularly attractive in a solar economy would be the use of biomass for the co-generation of electricity and industrial process steam.

While no single solar technology can meet humankind's total demand for energy, a combination of solar sources can. The transition to a solar era can be begun today; it would be technically feasible, economically sound, and environmentally attractive. Moreover, the most intriguing aspect of a solar transition might lie in its social and political ramifications.[10]

The kind of world that could develop around energy sources that are efficient, renewable, decentralized, simple, and safe cannot be fully visualized from our present vantage point. Indeed, one of the most attractive promises of such sources is a far greater flexibility in social design than is afforded by their alternatives. Although energy sources may not dictate the shape of society, they do limit its range of possibilities; and dispersed solar sources are more compatible than centralized technologies with social equity, freedom, and cultural pluralism. All in all, solar resources could power a rather attractive world.

Solar Heating

Solar energy is most easily captured as low-grade heat. Development of the flat-plate collector that is used to catch such heat is generally credited to the eighteenth-century Swiss scientist Nicolas de Saussure, who obtained temperatures over 87 degrees Centigrade using a simple wooden box with a black bottom and a glass top. The principle used by Saussure is simple: glass is transparent to sunlight but not to the radiation of longer wave lengths given off by the hot collector itself. Sunlight flows easily through the glass top into the collector, where it is trapped as heat. The modern flat-plate collector operates on this same basic principle, although improved materials achieve much higher temperatures and are more durable. Simple and easy-to-make solar collectors could supply heat now provided by high-quality fuels. More than one-third of the energy budget of all nations is spent to produce heat at temperatures that flat-plate solar collectors can achieve.[11]

The simplest task to accomplish directly with solar power is heating water, and solar water heaters are being utilized in many countries. More than two million have been sold in Japan, and tens of thousands are in use in Israel. In the remote reaches of northern Australia, where fuels are expensive, solar water heaters are required by law on all new buildings. Until replaced by cheap natural gas, solar water heaters were much used in California and in Florida; Miami alone had about 50,000 in the early 1950s. Since 1973, interest in solar water heaters has rekindled in many parts of the world. In poorer countries, cheap hot water can make a significant contribution to public well-being: hot water for dishwashing and bathing can reduce the burden of infectious diseases, and clothes washed with hot water and soap outlast clothes beaten clean on rocks at a river's edge.

Sunlight can also be used to heat buildings. All buildings receive and trap radiant energy from the sun. For warming a home on a winter day, this heat may be desirable; but it can constitute indecent exposure, broiling and embroiling the occupants of an all-glass office building, in midsummer. Solar buildings, designed to anticipate the amount of solar energy available in each season, put sunlight to work. To harness diffuse solar energy to meet a building's needs, options that vary in efficiency, elegance, and expense can be employed.[12]

Solar collectors are being used in diverse locations to heat buildings. The town of Mejannes-le-Clap in southern France has announced plans to obtain most of its heat from the sun. Several U.S. solar-heated communities, as well as individual schools, meeting halls, office buildings, and even hamburger stands, are now under construction. Saudi Arabia plans to build a new town at Jubail, using sunlight for heating, cooling, and for running water pumps; the Saudis are now also building the world's largest solar-heated building—a 325,000-square-foot athletic field house—in Tabuk.

In addition to warming buildings, low-grade heat from simple solar devices can also be used to dry crops—a task that now often consumes prodigious amounts of propane and methane gas. Solar dryers are now being used to remove moisture from lumber and textiles, as well as from corn, soybeans, alfalfa, raisins, and prunes. The sun has always been used to dry most of the world's laundry.

For more than a century, solar advocates have gathered crowds by cooking food with devices that use mirrors to intensify sunlight. Now that firewood supplies are growing scarce in many parts of the Third World, solar cooking is being taken more seriously. Although solar cookers proved popular in some village experiments in the 1960s, their high cost, as much as $25 each, prohibited widespread use. Today, however, cheap new reflecting materials like aluminized mylar can be stretched over inexpensive locally made frames. In poor countries, solar cookers will be only supplementary devices for now, since these mechanisms cannot function at night or in cloudy weather and since storing high-temperature heat is expensive. But if heat-storage technology advances, solar stoves and ovens may play an increasingly important role in rich and poor countries alike.

Solar technology now also encompasses desalination devices that evaporate water to separate it from salt. In the late nineteenth century, a huge solar desalination plant near Salinas, Chile, provided up to 6,000 gallons of fresh water per day for a nitrate mine. Recent research has led to major improvements in the technology of solar desalination, especially to improvements in "multiple-effect" solar stills. Today this sun-driven process holds great promise, especially in the Middle East and other arid regions. A small Soviet solar desalination plant in the Kyzyl-Kum Desert in central Asia now produces four tons of fresh water a day.[13]

Relatively low temperature sources of heat can also be used to operate pumps and engines. In the 1860s, Augustin Mouchot, a French physicist, developed a one-half-horsepower solar steam engine. In the early twentieth century, more efficient engines were built using ammonia and ether instead of water as the working fluid. In 1912, Frank Shuman constructed a 50-horsepower solar engine near Cairo to pump irrigation water from the Nile.

Scores of solar devices were built around the world in the early decades of this century, but none withstood the economic competition of low-cost fossil fuels. In recent years, with fuel prices soaring, solar pumps have begun to attract attention again. In 1975, a 40-horsepower solar pump of French design was installed in San Luis de la Paz to meet this Mexican town's irrigation and drinking needs. Mexico has ordered ten more such pumps; and Senegal, Niger, and Mauritania have installed similar devices. At present, solar pumps make economic sense only in remote areas where fuel and maintenance costs for conventional systems are extremely high. But, many authorities believe, the costs of solar pumps could be dramatically reduced by taking advantage of the findings of further research and the economies of mass production.[14]

Solar energy can be used directly in various industrial processes. A study of the Australian food-processing industry found, for example, that heat comprised 90 percent of the industry's energy needs; almost all this heat was at under 150 degrees Centigrade, and 80 percent was below 100 degrees. Such low-temperature heat can be easily produced and stored using elementary solar technologies. Similarly, a study of an Australian soft-drink plant found that enough collectors could be retrofitted onto the factory's roof to provide 70 percent of all the plant's heat requirements.[15]

A recent study of U.S. industrial heating demands concludes that about 7.5 percent is used at temperatures below 100 degrees Centigrade and 28 percent below 288 degrees. However, direct solar power can be used to preheat materials from ambient temperatures to intermediate temperatures before another energy source is employed to achieve the still higher temperature demanded for an industrial process. Such solar preheating can play a role in virtually every industrial heat application. If preheating is used, 27 percent of all energy for U.S. industrial heat can be delivered under 100 degrees Centigrade and about 52 percent under 288 degrees.[16]

Much of the energy used in the residential, commercial, agricultural, and industrial sectors is employed as low-temperature heat. In the recent past, this demand has been filled by burning fossil fuels at thousands of degrees or nuclear fuels at millions of degrees. Because such energy sources were comparatively cheap, little thought was given to the obvious thermodynamic inefficiency of using them for low-grade heat. Now that fuel costs are mounting rapidly, however, demands for heat will be increasingly met directly from the sun.

Electricity from the Sun

It was long believed that nuclear power would replace the fossil fuels. Because nuclear power is best utilized in centralized electrical power plants, virtually all energy projections therefore show electricity fulfilling a growing fraction of all projected energy demands. Some solar proponents advocate large centralized solar power plants as direct replacements for nuclear power plants to meet this demand. However, solar technologies can provide energy of any quality, and remarkably little of the world's work requires electricity. A sensible energy strategy demands more than the simpleminded substitution of sunlight for uranium.[17]

Electricity now comprises less than 20 percent of energy use in virtually all countries. If energy sources were carefully matched with energy uses, it is difficult to imagine a future society that would need more than one-tenth of its energy budget as electricity—the highest quality and most expensive form of energy. Today only 11 percent of U.S. energy use is electricity, and much of this could be met with other energy sources. To fill genuine needs for electricity, the most attractive technology in many parts of the world will be direct solar conversion.

Two types of large, land-based solar thermal power plants are receiving widespread attention. The "power tower" is currently attracting the most money and minds, although a rival concept—the "solar farm"— is also being investigated. The power tower relies upon a large field of mirrors to focus sunlight on a boiler located on a high structure—the "tower." The mirrors are adjusted to follow the sun across the sky, always maintaining an angle that reflects sunlight back to the boiler. The boiler, in turn, produces high-pressure steam to run a turbine to generate electricity. The French, who successfully fed electricity into their national grid from a small tower prototype in January of 1977, plan to have a 10-megawatt unit operating by 1981 and have been aggressively trying to interest the desert nations of the Middle East in this effort. The United States is now testing a small prototype involving a 40-acre mirror field and a 200-watt tower in New Mexico, and it plans to put a 10-megawatt power plant into operation by 1980 at Barstow, California.

An electric utility in New Mexico plans to combine three 430-foot power towers that generate a total of 50 megawatts with an existing gas-fired power plant at Albuquerque. The proposed complex would utilize the existing generators, turbines, condensers, switchyard, and so on. The resulting hybrid, which would cost $60 million and cover 170 acres, would have no heat storage capacity; it would simply heat its boilers with gas when the sun failed to shine. A survey by the utility identified 600 existing power plants in the American Southwest (with about 40,000 megawatts of electrical generating capacity) that could be retrofitted with solar power towers.[18]

The "solar farm" concept would employ rows of parabolic reflectors to direct concentrated sunlight onto pipes containing molten salts or hot gases. Special heat exchangers would transfer the 600-degree Centigrade heat from the pipes to storage tanks, filled with melted metal, from whence it could be drawn to generate high-pressure steam to run a turbine.

Both the solar farm and the power tower approaches required direct sunlight because their concentrating mirrors cannot use diffuse light. Both will also probably be feasible only in semi-arid regions with few cloudy days and little pollution. One objection raised to such facilities is that they would despoil large tracts of pristine desert. However, proponents point out that the area needed to produce 1,000 megawatts of solar electricity is less than the amount of land that would have to be strip-mined to provide fuel for a similar-sized coal plant during its thirty-year lifetime and that the solar plant's land could be used forever. In fact, according to Aden and Marjorie Meinel, a 1,000-megawatt solar farm on the Arizona desert would require no more land than must, for safety reasons, be deeded to a nuclear reactor of the same capacity.[19]

Large, centralized solar electric plants consume no finite fuels, produce no nuclear explosives, and hold no ecological punches. With development, such plants should also be economically competitive with fossil fueled, fission, and fusion power plants. However, they produce only electricity and they are subject to all the problems inherent in centralized high technologies. To the extent that energy needs can be met with lower quality sources or decentralized equipment, the centralized options should be avoided.

As a power source in countries where land is scarce or where cloud cover is frequent, solar electric plants are less promising; efficient long-distance cryogenic electrical transmission may prove technically feasible but will probably be extremely expensive. Proposals to tap North African deserts for power for Western Europe or to course Mexican sunlight through New York's power grid are therefore unlikely to bear fruit. A more likely consequence of solar thermal-electric development would be the relocation of many energy-intensive industries in sunny climes. In fact, Professor Ignacy Sachs, director of the International Center for Research on Environment and Development in Paris, has predicted that a new solar-powered industrial civilization will emerge in the tropics.

Land-based solar electric plants must bow to one incontrovertible fact: it is always night over half the earth. If such facilities are to generate power after the sun sets, oversized collectors must be built and the excess heat retained in an expensive storage facility until it is needed. But ocean thermal-electric conversion (OTEC) plants, which use the ocean as a free collector and storage system, are unaffected by daily cycles. Because the ocean's temperature varies little, OTEC plants can be a source of steady, round-the-clock power.

The temperature difference between the warm surface waters of tropical oceans and the colder waters in the depths is about 20 degrees Centigrade. In 1881, J. D'Arsonval suggested in an article in *Revue scientifique* that this difference could be used to run a closed-cycle engine. In the 1920s, another French scientist, Georges Claude, persuaded the French government to build a number of open-cycle power plants to exploit these ocean thermal gradients. After World War II, the French government built several OTEC plants (the largest of which had a capacity of 7.5 megawatts) in the hope that such plants would provide inexpensive energy to France's tropical colonies. French interest in the project crumbled along with its overseas empire, but the idea of harnessing ocean thermal gradients to generate power lingers on.[20]

Because of the small temperature differences between deep and surface waters, OTEC's potential efficiency is severely limited. Moreover, as much as a third of the power an OTEC facility produces may be required to pump the enormous amounts of water needed to drive the cycle. Despite these difficulties and the additional problem of transporting power to users on the shore, OTEC proponents contend that the system will be cheap enough to underprice competing sources of electricity. However, this contention is untested, and estimates of an OTEC unit's cost range from about $450 to almost $4,000 per installed kilowatt—excluding the costs of transporting the electricity to the land and the costs of any environmental damages. The real cost will probably fall between these extremes, but early models, at least, will likely veer toward the high end.[21]

The OTEC concept does not involve any new basic technology. Its proponents tend to downplay the technical difficulties as simply matters of "good plumbing," even though the system would require pumps and heat exchangers far larger than any in existence. Because they do not consume any fuel, OTEC systems are largely insured against future cost increases that could affect nuclear or fossil fueled plants. On the other hand, with so much of their costs as, literally, sunk investments, the viability of OTECs will depend entirely upon their durability and reliability—two open questions at this point. Unexpected vulnerabilities to corrosion, biological fouling, hurricanes, or various other plagues could drive costs up dramatically.

Intensive deployment on the scale urged by OTEC's most ardent advocates could also possibly engender a variety of environmental problems that a few scattered plants would not provoke. An increase in the over-all heat of substantial bodies of water and the upwelling of nutrient-rich waters from the ocean bottom could both provoke unfortunate consequences. Ocean temperature shifts could have far-reaching impacts on weather and climate, and displacing deep waters would disturb marine ecology. In addition, physicist Robert Williams of Princeton calculates, the upwelling of carbon-rich water from the ocean bottom could cause atmospheric carbon dioxide to increase substantially.[22] OTECs, like other large centralized sources of electricity, have costs that multiply rapidly when large numbers of plants are built. This technology should probably be limited to a modest number of facilities in ocean areas where conditions are optimal.

The most exciting solar electric prospect is the photovoltaic cell— now the principal power source of space satellites. Such cells generate electricity directly when sunlight falls on them. They have no moving parts, consume no fuel, produce no pollution, operate at environmental temperatures, have long lifetimes, require little maintenance, and can be fashioned from silicon, the second most abundant element in the earth's crust.[23]

Photovoltaic cells are modular by nature, and little is to be gained by grouping large masses of cells at a single collection site. On the contrary, the technology is most sensibly applied in a decentralized fashion—perhaps incorporated in the roofs of buildings—so that transmission and storage problems can be minimized. With decentralized use, the 80 percent or more of the sunlight that such cells do not convert into electricity can also be harnessed to provide energy for space heating and cooling, water heating, and refrigeration.

Fundamental physical constraints limit the theoretical efficiency of photovoltaic cells to under 25 percent. Numerous practical problems force the real efficiency lower—for silicon photovoltaics, the efficiency ceiling is about 20 percent. To obtain maximum efficiency, relatively pure materials with regular crystal structures are required. Such near perfection is difficult and expensive to obtain. High costs have, in fact, been the principal deterrent to widespread use of photovoltaic cells.[24]

Cost comparisons between photovoltaic systems and conventional systems can be complicated. Solar cells produce electricity only when the sun shines; conventional power plants, on the other hand, are frequently shut down for repairs or maintenance. Depending on the amount of sunlight available where a photovoltaic array is located, the cells might produce between one-fourth and one-half as much power per kilowatt of installed capacity as an average nuclear power plant does. Adding to the costs of photovoltaics is the need for some kind of storage system; on the other hand, the use of photovoltaics may eliminate the need for expensive transmission and distribution systems.

Depending upon who does the figuring, photovoltaic cells now cost between twenty and forty times as much as conventional sources of base-load electricity. However, as a source of power just during daylight periods of peak demand, photovoltaics cost only four to five times as much as conventional power plants plus distribution systems. Moreover, the costs of conventional power plants have shot steadily upward in recent years while the costs of photovoltaic cells have rapidly declined, and several new approaches are being pursued in an effort to further diminish the costs of photovoltaic arrays. For example, focusing collectors that use inexpensive lenses or mirrors to gather sunlight from a broad area and concentrate it on the cells are being employed. The Winston collector can obtain an eight-to-one concentration ratio without tracking the sun; "tracking" collectors can obtain much higher multiples, but at far greater expense.

Another approach to cutting the costs of photovoltaic cells has been to use less efficient but much cheaper materials than those usually used; amorphous silicon and a combination of cadmium sulfide and copper sulfide are strong candidates. Although the required collector area is thus increased, total costs may be less. Conversely, another approach has been to improve the processing of high-grade materials for photovoltaic cells. Currently, each cell is handcrafted by artisans who use techniques not unlike those employed in a Swiss watch factory. Simple mechanization of this process could lead to large savings. The costs of photovoltaic cells, which amounted to $200,000 per peak kilowatt in 1959, have already fallen to about $13,000 per peak kilowatt and most experts believe that prices will continue to fall rapidly.[25]

Increased production is of paramount importance in lowering the prices of photovoltaics. In an eighteen-month period of 1975–76, U.S. purchases of photovoltaic cells for earthbound purposes doubled and the average price per cell dropped by about 50 percent. Price reductions of from 10 to 30 percent for each doubling of output have been common in the electrical components industries, and photovoltaic production should prove no exception to the rule.

The objective of the Low-Cost Silicon Array Project of the U.S. Energy Research and Development Administration is to produce photovoltaics for less than $500 per peak kilowatt, and to be annually producing more than 500 megawatts by 1985. This program, contracted through the California Institute of Technology, involves a large number of major corporations. A general consensus appears to be developing among the participants that the goals are reachable and may even be far too modest. Under the auspices of the government's "Project Sunshine," Japan has undertaken a similar research effort.[26]

From a "net energy" perspective, photovoltaics are appealing. Detailed studies of the energy needed to manufacture such cells shows that the energy debt can be paid in less than two years of operation. With more energy-efficient production processes, the energy payback period could, theoretically, be reduced to a matter of weeks. If the energy some cells produce is fed back to produce more cells, photovoltaics can become true energy "breeders"—making more and more energy available each year without consuming any nonrenewable resources. In fact, Malcolm Slesser and Ian Hounam have calculated, an initial 1-megawatt investment in photovoltaic cells with a two-year payback period could multiply in forty years to provide 90 percent of the world's energy needs. These calculations may be a bit optimistic, and the world does not want or need to consume 90 percent of its energy in the form of electricity; but photovoltaics, like other solar technologies, hold up well under net energy analysis.[27]

A variety of options are available to produce electricity directly from the sun. Several of the approaches sketched here—all of which have been technically demonstrated—are now economically competitive with fossil fueled plants under some conditions. Prices can be reasonably expected to fall dramatically as more experience is gained. Although solar electricity will probably never be really cheap, it is doubtless worth paying some economic premium for a source of electricity that is safe, dependable, renewable, non-polluting, and—in the case of photovoltaics—highly decentralized.[28]

Storing Sunlight

Jets and trucks cannot run directly on sunbeams. At night, of course, nothing can. Solar energy is too diffuse, intermittent, and seasonally variable to harness directly to serve some human needs. Of course, interruptions of various kinds plague all energy systems, and storage problems are not unique to renewable power sources. Electrical power lines snap, gas and oil pipelines crack, dams run low during droughts, and nuclear power plants frequently need repairs and maintenance. A wind turbine on a good site with sufficient storage capacity to handle a ten-hour lull could, Danish physicist Bent Sørensen has shown, deliver power as reliably as a typical modern nuclear power plant. Reliability is thus a relative concept.[29]

Sometimes the intermittent nature of an energy source causes no problems. For example, solar electric facilities with no storage capacity can be used to meet peak demands, since virtually all areas have their peak electrical demands during daylight hours. Some users, such as fertilizer producers, may find that an intermittent energy source satisfies their needs. And sometimes two intermittent sources will complement each other. For example, wind speeds are usually highest when the sun is not shining, so wind and solar devices can often be effectively used in tandem.

Often, however, energy must be stored. One option is to store energy as heat. Low-temperature heat for warming buildings, for example, can be temporarily stored in such substances as water or gravel; in fact, substantial short-term heat storage capacity can be economically designed into the structural mass of new buildings. For longer periods, eutectic (phase-changing) salts are a compact, effective storage medium. Higher temperature heat, suitable for generating electricity, can be stored in hot oil or perhaps in melted sodium. A 1976 report for the U.S. Electric Power Research Institute rated thermal storage (along with pumped hydro-storage and compressed air storage) as the most promising options for centralized utilities.[30]

Many solar enthusiasts are intrigued by hydrogen storage systems. The distinguished British scientist and writer J. B. S. Haldane predicted in a lecture given at Cambridge University in 1923 that England would eventually turn for energy to "rows of metallic windmills working electric motors." Haldane then went on:

At suitable distances, there will be great power stations where during windy weather the surplus power will be used for the electrolytic decomposition of water into oxygen and hydrogen. These gases will be liquefied and stored in vast vacuum jacketed reservoirs, probably sunk in the ground. . . . In times of calm, the gases will be re-combined in explosion motors working dynamos which produce electrical energy once more, or more probably in oxidation cells.[31]

Little has been done to advance large-scale hydrogen usage since Haldane startled Cambridge with his vision more than a half century ago. The reason for the pause is easy enough to fathom; fossil fuels were for decades so cheap that hydrogen could not be made competitive. In recent years, interest in hydrogen has revived, partly because this fuel has been used so successfully in space exploration programs and partly because natural gas companies have gradually begun to awaken from their "pipe dreams" of endless natural gas supplies.

Under some grand schemes, hydrogen would someday substitute for all natural gas, replace all automobile fuel, and satisfy much of industry's total energy demand as well. But the most farfetched of such plans for a "hydrogen economy" strain the imagination. The easiest way to make hydrogen (other than by reforming fossil hydrocarbons) is by electrolyzing water; the United States would have to triple its present electrical generating capacity in order to substitute hydrogen for the natural gas it now uses—even if it were to devote *all* its electricity to the task.

Hydrogen production poses a technical problem but it is one that may eventually yield to a cheap technical solution. In fact, some promising research is now being conducted on biological production processes and on techniques for using high-temperature solar heat to split water molecules into hydrogen and oxygen. In the meantime, hydrogen recommends itself for use in storing and transporting energy from intermittent sources of power. Easily stored as a pressurized gas, as a super-cooled liquid, or in metal hydrides, hydrogen can also be transported long distances more economically than electricity and can be used in fuel cells (where it can be efficiently converted into electricity in decentralized facilities). Pressurized hydrogen tends to embrittle some metals and alloys, but the importance of this problem has probably been exaggerated.[32]

Pumped hydro-storage involves using surplus power to pump water from a lower reservoir to an elevated one. Then, when power is needed, the water is allowed to flow back to the lower pool through a turbine. Pumped hydro-storage is already used with conventional power plants around the world; in the future it may be crossbred with wind-power technologies. The use of wind energy declined in Denmark a half century ago in part because "wind muscle" could not compete economically with cheap, surplus Swedish hydropower. Now that demand for electricity has increased in both countries, both are seriously considering investing in a hybrid system. Danish wind power could replace some Swedish hydropower when the wind blows, and any surplus wind power could be used to pump downstream water back into some of Sweden's reservoirs. Sweden might also pursue wind power independently. The Swedish State Power Board has determined that 5,000 megawatts of wind-power capacity could be linked with current hydroelectric facilities without providing extra storage. Such a combination of wind power and hydropower would make sense in many places: when a dam has excess capacity and could generate more electricity without adding more turbines if only it held more water, a hybrid system fits the bill. The Bonneville Power Administration is considering the integration of wind turbines into its extensive hydroelectric system in the northwestern United States.

Another form of mechanical storage involves pumping pressurized air into natural reservoirs (e.g., depleted oil and gas fields), man-made caverns (including abandoned mines), or smaller specially made storage tanks. Air stored in this manner is released as needed to drive turbines or to run machinery. For almost four decades, designers have studied large-scale pumped-air storage proposals, but the first commercial unit is just being completed. Located in Huntorf, West Germany, the system will store the surplus power generated by nuclear reactors during periods of low power demand.[33]

Still another approach to mechanical storage involves rapidly rotating flywheels in environments that are almost friction-free. Recent major advances in materials now allow the construction of "super-flywheels" whose higher spinning speeds enable them to store large amounts of energy in rather small areas. Flywheels could, in theory, be made small and efficient enough to propel individual automobiles. They have already been used in pilot projects on trolleys and buses to recapture the energy that would otherwise be lost during braking. Although superflywheels seem attractive at first blush, significant problems remain; and these devices are some years away from widespread commercial application.[34]

Electricity can be stored directly in batteries. Existing batteries are rather expensive, have low power and energy densities, and do not last long. However, experimental batteries, some of which may prove economical and feasible when used with intermittent energy sources, may soon enter the market. Metal-gas batteries, like the zinc chloride cell, use inexpensive materials and have relatively high energy densities. Alkali-metal batteries perform very well, but operate at high temperatures, and existing models suffer from short life spans. A number of other battery possibilities are being investigated and some promising preliminary research results are now emerging.[35]

Base-load sources of electricity, such as coal plants and nuclear plants, also require storage. Such facilities cannot be geared up and down to follow the peaks and valleys of electrical usage; they produce power at a steady rate, and surplus power from non-peak hours must be stored for the periods of heaviest demand. For base-load plants, the cost of storage varies with the degree to which consumer usage is not constant twenty-four hours a day. For solar sources, the storage costs vary with the extent to which usage does not coincide with the normal daytime sunlight cycle. Wind power is less predictable but at choice sites tends to be quite constant. Storage problems with hydropower and biomass systems are minimal. Over all, the storage requirements for a society based on renewable energy sources may prove comparable to those of an all-nuclear society.

Sunlight is abundant, dependable, and free. With some minor fluctuations, the sun has been bestowing its bounty on the earth for more than four billion years, and it is expected to continue to do so for several billion more. The sun's inconstancy is regional and seasonal, not arbitrary or political, and it can therefore be anticipated and planned for.[36]

Different solar sources will see their fullest development in different regions. Wind-power potential is greatest in the temperate zones, while biomass flourishes in the tropics. Direct sunlight is most intense in the cloudless desert, while water power depends upon mountain rains. However, most countries have some potential to harness all these renewable resources, and many lands have begun to take advantage of some of them. In many small ways in many diverse places, the solar transition has already begun.

10. Wind and Water Power

Catching the Wind

THE AIR that envelops the earth functions as a 20-billion-cubic-kilometer storage battery for solar energy. Winds are generated by the uneven heating of our spinning planet's land and water, plains and mountains, equatorial regions and poles. The idea of harnessing this wind to serve human needs may have first occurred to someone watching a leaf skitter across a pond. Five thousand years ago, the Egyptians were already sailing barges along the Nile. Wind-powered vessels of one sort or another dominated shipping until the nineteenth century, when ships driven by fossil fuels gradually eased them out. A few large cargo schooners plied the east coast of the United States until the 1930s, and the largest windjammers were the greatest wind machines the world has known.[1]

The windmill appears to have originated in Persia two millennia ago. There, vertical shaft devices that turned like merry-go-rounds were used to grind grain and pump water. After the Arab conquest of Persia, wind power spread with Islam throughout the Middle East and to the southern Mediterranean lands. Invading Mongols carried the windmill back to China. Returning Crusaders likewise appear to have transferred the technology to Europe—though the tilt (30 degrees to the horizontal) of the axes of early European mills have led some scientists to believe that the device may have been invented independently by a European. Eventually, horizontal-axis windmills with blades that turned like Ferris wheels were developed, and they spread throughout Europe.[2]

By the seventeenth century, the Dutch had a commanding lead in wind technology and were already using wind power to saw wood and make paper. In the late nineteenth century, the mantle of leadership passed to the Danes, who had about 100,000 windmills in operation by 1900. Under the leadership of Poul la Cour, Denmark began making significant investments in wind-generated electricity and by 1916 was operating more than 1,300 wind generators.

The windmill played an important role in American history, especially in the Great Plains, where it was used to pump water. More than six million windmills were built in the United States over the last century; about 150,000 still spin productively. Before the large-scale federal commitment to rural electrification in the 1930s and 1940s, windmills supplied much of rural America with its only source of electricity.

After World War I, cheap hydropower and dependable fossil fuels underpriced wind-power plants. However, research in many parts of the world continued, and many interesting windmill prototypes were constructed. In 1931, the Soviet Union built the world's first large wind generator near Yalta. Overlooking the Black Sea, this 100-kilowatt turbine produced about 280,000 kilowatt-hours of electricity per year. In the 1950s, Great Britain built two 100-kilowatt turbines. In 1957, Denmark built a 200-kilowatt turbine, and France constructed an 800-kilowatt wind generator. In 1963, a 1,000-kilowatt wind turbine was built in France.

The largest wind generator ever built was the 1,250-kilowatt Grandpa's Knob machine designed by Palmer Putnam and erected on a mountaintop in central Vermont. It began generating electricity on August 29, 1941, just two years after its conception. However, the manufacturer had been forced to cut corners in his haste to finish construction before the icy hand of wartime rationing closed upon the project, and the eight-ton propeller blades developed stress cracks around their rivet holes. Although the cracking was noticed early, the blades could not be replaced because of materials shortages. Finally, a blade split, spun 750 feet in the air, and brought the experiment to a crashing conclusion. The private manufacturer had invested more than a million dollars in the project and could afford to risk no more.[3]

Despite the enthusiasm of occasional wind-power champions in the federal government, no more major wind generators were constructed in the United States until 1975. Then NASA began operating a 100-kilowatt prototype near Sandusky, Ohio, that resembles a huge helicopter mounted sideways atop a transmission tower. The next major step in the American program will be a 1,500-kilowatt wind turbine to be built jointly by General Electric and United Technology Corporation in 1978.

Before the GE-UTC turbine begins operating, however, it may have slipped into second place in the size sweepstakes. Tvind, a Danish college, has nearly completed a 2,000-kilowatt wind turbine, at a cost of only $350,000. (Doubtless the most important factor in holding down expenses for the Tvind generator is that the college staff paid for the project out of their own pockets. If successful, Tvind will hearten those who hope that major technical accomplishments can still be achieved without reliance on central governments or big business.)[4]

The Tvind wind machine, like virtually all large wind turbines today, will have only two blades. While more blades provide more torque in low-speed winds (making multiple blades particularly useful for purposes such as small-scale water pumping), fewer blades capture more energy for their cost in faster winds. A two-blade propeller can extract most of the available energy from a large vertical area without filling the area with metal that could crack or split in a storm.

Since power production increases with the square of a turbine's size, large wind machines produce far more energy than do small ones. Moreover, wind power increases as the cube of velocity, so a 10-meter-per-second wind produces eight times as much power as a 5-meter-per-second breeze does. Consequently, some wind-power enthusiasts limit their dreams to huge turbines on very windy sites. In particular, a recent survey of large U.S. corporations conducting wind-power research disclosed that only one company had any interest in small or intermediate-sized turbines.[5]

However, the "think big" approach does not necessarily make sense. The crucial question for windmills is how much energy is harnessed per dollar of investment. Increases in output are desirable only if the value of the additional energy extracted exceeds the extra cost, and economic optimization does not necessarily lead to the construction of giant turbines. Smaller windmills might lend themselves more easily to mass production and might be easier to locate close to the end-user (thus reducing transmission costs). Small windmills can produce power in much lower winds than large ones do and can thus operate more over a given time. Smaller-scale equipment also allows a greater decentralization of ownership and control, and the consequences of equipment failure are not likely to be catastrophic. Finally, wind turbine development will probably be constrained by practical limits on propeller size. Large turbines place great stresses on both the blade and tower, and all giant turbines built to date have suffered from metal fatigue.

On a small scale, wind power can be cheaply harnessed to perform many kinds of work. The Valley of Lasithi on Crete uses an estimated 10,000 windmills, which catch the wind in triangular bands of white sailcloth, to pump irrigation water. Similar windmills built of local materials have recently been erected in East Africa. The New Alchemy Institute in Massachusetts, working with the Indian Institute of Agricultural Research and the Indian National Aeronautical Laboratory, has developed a 25-foot sail-wing pump for rural use; employing the wheel of a bullock cart as the hub and a bamboo frame for the cloth sails, this simple machine could provide cheap power to Indian villages. The Brace Research Institute in Canada has designed a Savonius water pump that can be constructed from two 45-gallon oil drums cut in half. Already used in the Caribbean, the device costs about $50 to make and will operate at wind speeds as low as 8 mph.

Traditionally, wind has been used primarily to pump water and to grind grain. Windmills can also produce heat that can be stored and used later in space heating, crop drying, or manufacturing processes. A particularly attractive new approach is to compress air with wind turbines. Pressurized air can be stored much more easily than electricity, a fact to which virtually every gasoline station in the United States attests. Stored air can either be used as needed to directly power mechanical equipment or released through a turbine to generate electricity. On a large scale, pressurized air can be stored in underground caverns.

The modern wind enthusiast can choose from many options: multiple-blade propellers, triple-blade props, double-blade props, single-blade versions with counterweights, sail wings, crosswind paddles, and gyromills. In some wind turbines, the propeller is upwind from the platform, while in others it is located downwind. Some platforms support single large turbines; others support many small ones. A machine with two sets of blades turning in opposite directions is being tested in West Germany.[6]

One of the most interesting multiple-blade devices for small and moderate-sized generators is under development at Oklahoma State University. This mill resembles a huge bicycle tire, with flat aluminum blades radiating out from the hub like so many spokes. Instead of the generator's being geared to the hub of the windmill, the Oklahoma State machine operates on the principle of the spinning wheel: the generator is connected to a belt that encircles the faster-moving outer rim.

The Darrieus wind generator, favored by the National Research Council of Canada and by Sandia Laboratories in New Mexico, looks like an upside-down egg beater, and turns around its vertical axis like a spinning coin. The Darrieus holds several striking advantages over horizontal-axis turbines: it will rotate regardless of wind direction; it does not require blade adjustments for different wind speeds; and it can operate without an expensive tower to provide rotor clearance from the ground. Aerodynamically efficient and lightweight, the Darrieus might cost as little as one-sixth as much as a horizontal-shaft windmill of the same capacity. In early 1977, a 200-kilowatt Canadian Darrieus wind turbine began feeding electricity into the 24,000-kilowatt power grid that serves the Magdalen Islands in the Gulf of St. Lawrence. If this machine lives up to its economic potential, other Darrieus turbines will be installed.[7]

Intriguing new approaches to wind power may well be gestating. Little money or effort has been put into wind turbine research over the last two decades, although aeronautical engineering has made enormous strides over the same period. With interest in wind machines gathering force, new approaches might well emerge. For example, a "confined vortex" generator being developed by James Yen steers wind through a circular tower, creating a small tornado-like effect; this generator utilizes the difference in pressure between the center of the swirling wind and the outside air to drive a turbine.[8] Large amounts of electricity could, theoretically, be generated by relatively small turbines of this type. The U.S. Energy Research and Development Administration recently awarded Dr. Yen $200,000 to develop this idea further. However, the viability of wind power does not depend upon scientific breakthroughs; existing wind technologies can compete on their own terms for a substantial share of the world's future energy budget.

Estimating the probable cost of wind power is a somewhat speculative undertaking. The cost of generating electricity with the wind can be measured in two different ways—depending upon whether the system provides "base load" power or only supplementary power. If wind generators feed power directly into a grid when the wind blows, and if other generating facilities have to be constructed to handle peak loads when the wind isn't blowing, the average costs of building and maintaining such windmills must compete with just the cost of fuel for the alternative power plant. Obviously, this calculation hinges not only on how much a windmill costs to construct, but also on how long it lasts and how reliably it functions. Conclusions are premature until experience has been gained, but many studies have suggested that intermittent electricity could be generated today from the wind for considerably less than the cost of merely providing fuel for an oil-fired unit. Moreover, wind-power costs could diminish significantly as more experience is acquired, while oil costs will certainly rise.[9]

If the wind is used to provide constant, reliable power, then the cost of building a wind generator plus a storage facility must not exceed the total cost (including the environmental cost) of building and operating a conventional power plant. Used in conjunction with a hydroelectric facility with reserve capacity, wind turbines should already have a substantial cost advantage over conventional power plants. For other storage setups, cost calculations remain unsubstantiated, but studies of analogous technologies suggest that such base-load wind systems will be economically sound. When social and environmental costs are included, the case becomes even stronger. Accordingly, such systems should now be built and operated so that these calculations can be proven.

The rate and extent to which wind power is put to work is much more likely to be a function of political considerations than of technical or economic limits. The World Meteorological Organization has estimated that 20 million megawatts of wind power can be commercially tapped at the choicest sites around the world, not including the possible contributions from large clusters of windmills at sea.[10] By comparison, the current total world electrical generating capacity is about 1.5 million megawatts. Even allowing for the intermittent nature of the resource, wind availability will not limit wind-power development. Long before a large fraction of the wind's power is reaped, capital constraints and social objections will impose limits on the growth of wind power.

Well-designed, well-placed wind turbines will achieve a high net energy output with an exceptionally mild environmental and climatic impact: wind machines produce no pollution, no hazardous materials, and little noise. In fact, the principal environmental consequences of wind power will be comparatively modest ones associated with mining and refining the metals needed for wind turbine construction—ill-effects associated with virtually every energy source. Windmills will have to be kept out of the migratory flyways of birds, but these routes are well known and can be easily avoided. Where objections to wind technology on aesthetic grounds arise, windmills could be located out of the visual range of populated areas, even a few miles out to sea. Moreover, some wind machines, such as the Darrieus, strike many as handsome. All things considered, a cleaner, safer, less disruptive source of energy is hard to imagine.[11]

Falling Water

Numerous surveys of the world's water-power resources suggest that a potential of about 3 million megawatts exists, of which about one-tenth is now developed. The figure is unrealistic, however, since reaching the 3-million-megawatt potential would require flooding fertile agricultural bottomlands and rich natural ecosystems. On the other hand, none of the surveys include the world's vast assortment of small hydroelectric sites. By even the most conservative standards, potential hydropower developments definitely exceed 1 million megawatts, while current world hydroelectric capacity is only 340,000 megawatts.

Industrialized regions contain about 30 percent of the world's hydroelectric potential as measured by conventional criteria but produce about 80 percent of all its hydroelectricity. North America produces about one-third, Europe just a little less, and the Soviet Union about one-tenth. Japan, with only 1 percent of the world's potential, produced over 6 percent of all its hydroelectricity. In contrast, Africa is blessed with 22 percent of all hydroelectric potential, but produces only 2 percent of all hydroelectricity—half of which comes from the Aswan High Dam in Egypt, the Akosombo Dam in Ghana, and the Kariba

Dam on the Zambezi River between Zambia and Rhodesia. Asia (excluding Japan and the USSR) has 27 percent of the potential resources, and currently generates about 12 percent of the world's hydroelectricity; most of its potential lies in the streams that drain the Tibetan Plateau, at sites far from existing energy markets. Latin America, with about 20 percent of the world's total water-power resources, contributes about 6 percent of the current world output. Nine of the world's fifteen most powerful rivers are in Asia, three are in South America, two are in North America, and one is in Africa.[12]

The amount of hydropower available in a body of moving water is determined by the volume of water and by the distance the water falls. A small amount of water dropping from a great height can produce as much power as a large amount of water falling a shorter distance. The Amazon carries five times as much water to the sea as does the world's second largest river, the Congo; but because of the more favorable topography of its basin, the Congo has more hydroelectric potential. In mountainous headwater areas, such as Nepal, where relatively small volumes of water fall great distances, numerous choice sites exist for stations of up to 100 megawatts each.[13]

Used by the Romans to grind grain, waterwheels reached their highest pre-electric form in the mid-1700s with the development of the turbine wheel. The Versailles waterworks produced about 56 kilowatts of mechanical power in the eighteenth century. In 1882, the first small hydroelectric facility began producing 125 kilowatts of electricity in Appleton, Wisconsin, and by 1925 hydropower accounted for 40 percent of the world's electric power. Although hydroelectric capacity has since grown fifteenfold, its share of the world's electricity market has fallen to about 23 percent.

Early hydroelectric development tended to involve small facilities in mountainous regions. In the 1930s, emphasis shifted to major dams and reservoirs in the middle and lower sections of a river, such as the Tennessee Valley dams in the United States and the Volga River dams in the Soviet Union. The world now has 64 hydroplants with capacities of 1,000 megawatts or more each: the Soviet Union has 16, the United States has 12, Canada has 11 (the United States and Canada share another), and Brazil has 10.

The environmental and social problems associated with huge dams and reservoirs far outweigh those surrounding small-scale installations or projects that use river diversion techniques.[14] Moreover, the increments by which small facilities boost a region's power supply are manageable. In contrast, a tripling or quadrupling of a power supply in one fell swoop by a giant dam can lead to a desperate search for energy-intensive industries to purchase the surplus, dramatically upsetting the politics and culture of an area.

Much of the extensive hydroelectric development in Japan, Switzerland, and Sweden has entailed use of comparatively small facilities, and such small units hold continuing promise for developing countries. In late 1975, China reportedly had 60,000 small facilities that together generated over 2 million kilowatts—about 20 percent of China's total hydroelectric capacity. The Chinese facilities are located in sparsely populated areas, thereby neutralizing the prohibitive transmission costs of sending electricity from huge centralized facilities. Local workers build the small earth-filled or rock-filled dams that provide substantial flood control and irrigation benefits as they bring power to the people.[15]

Nevertheless, building enormous facilities to capture as much power as possible while taking advantage of the economies of large scale is tempting. Although this approach has been used extensively and rather successfully in the temperate zone, many of the remaining prime locations are in the tropics, where troubles may arise. The Congo, for example, with a flow of 40,000 cubic meters per second and a drop of nearly 300 meters in the final 200 kilometers of its journey to the ocean, has an underdeveloped hydroelectric potential of 30,000 megawatts. But experience in other warm areas indicates that great care must be taken in exploiting such resources.

The Aswan High Dam provides a textbook case of the problems that can encumber a major hydroelectric development in the tropics. So trouble-ridden is Aswan that its costs largely offset its benefits. Although Aswan is a source of electricity, of flood and drought control, and of irrigation, the dam's users and uses sometimes conflict. For example, Aswan provides more than 50 percent of Egypt's electrical power, but its production is highly seasonal; during winter months, the flow of water through the dam is severely diminished while irrigation canals are cleaned. This reduced flow causes power generation to drop from a designed capacity of 2,000 megawatts to a mere 700 megawatts. Furthermore, lack of money for an extensive transmission grid has meant that electricity does not reach many of the rural villages that had hoped to benefit from the project.

Aswan saved Egypt's rice and cotton crops during the droughts in northeastern Africa in 1972 and 1973. Irrigation has increased food production by bringing approximately 750,000 formerly barren acres under cultivation, and by allowing farmers to plant multiple crops on a million acres that had previously been harvested only once a year. These timely boosts have enabled Egypt's food production to keep pace, though just barely, with its rapidly growing population. On the other hand, the dam has halted the natural flow of nutrient-rich silt, leaving downstream farmers to rely increasingly upon energy-intensive chemical fertilizers; and the newly irrigated areas are so plagued by waterlogging and mounting soil salinity that a $30 million drainage program is now needed. In addition, the canals in some areas rapidly clog with fast-growing water hyacinths.

The Aswan has also given a new lease to an age-old health hazard in Egypt. Schistosomiasis, a disease caused by parasitic worms carried by water snails, has long been endemic in the Nile delta where most of Egypt's population is concentrated, but in the past it was rarely found in upstream areas. Since the construction of the large dam, infestations of this chronic and debilitating affliction are also common along the Nile and its irrigation capillaries in Upper Egypt. Many of the major problems associated with Aswan should have been anticipated and avoided. Even now, Aswan's worst problems probably can be either solved or managed. But after-the-fact remedies will be costlier and less effective than a modest preventive effort would have been.

The inevitable siltation of reservoirs does more to spoil the use of dams as renewable energy sources than does any other problem. Siltation is a complex phenomenon that hinges upon several factors, one of which is the size of the reservoir. For example, the Tarbela Reservoir in Pakistan holds only about one-seventh the annual flow of the Indus, while Lake Mead on the Colorado can retain two years' flow. The life expectancy of the Tarbela is measurable in decades; Lake Mead will last for centuries. The rate of natural erosion, another factor in siltation, is determined primarily by the local terrain. Some large dams in stable terrains have a life expectancy of thousands of years; others have been known to lose virtually their entire storage capacity during one bad storm. Logging and farming can greatly accelerate natural erosion too; many reservoirs will fill with silt during one-fourth their expected life spans because these and other human activities ruin their watersheds.

Siltation, which affects the dam's storage capacity but not its power-generating capacity, can be minimized. Water can be sluiced periodically through gates in the dam, carrying with it some of the accumulated silt. Reservoirs can be dredged, though at astronomical costs. By far the most effective technique for handling siltation is lowering the rate of upstream erosion through reforestation projects and enlightened land use.[16]

Dams cannot be evaluated apart from their interaction with many other natural and artificial systems. They are just one component, albeit a vital one, of river basin management. Locks will have to be provided on navigable rivers, and fish ladders (one of the earliest victories of environmentalists) must be installed where dams block the spawning routes of anadromous fish. If a dam is located in a dry area, power generation must be coordinated with downstream irrigation needs. If a populated basin is to be flooded, the many needs of displaced people as well as the loss of fertile bottomland must be taken into account. Unpopulated basins are politically easier to dam, but in unsettled areas care must nevertheless be taken to preserve unique ecosystems and other irreplaceable resources.

Dams are vulnerable to natural forces, human error, and acts of war. The 1976 collapses of the Bolan Dam in Pakistan, the Teton Dam in Idaho, and a large earthen dam outside La Paz on Mexico's Baja Peninsula serve as emphatic reminders of the need for careful geological studies and the highest standards of construction.

Dams recommend themselves over most other energy sources. They provide many benefits unconnected to power production; they are clean; and their use does not entail the storage problems that plague so many other renewable sources. Indeed, using dams as storage mechanisms may be the most effective way to fill the gaps left by solar and wind power. In addition, the conversion of water power into electrical power is highly efficient—85 percent or more. Finally, dams can be instruments of economic equity; the greatest potential for future hydropower development lies in those lands that are currently most starved for energy.

Turning the Tides

Like wind power and hydropower, tidal power was first harnessed to mill grain. English tide mills built at Bromley-by-Bow in 1100 and at Woodbridge in 1170 functioned successfully for eight hundred years. Tide mills were built in Zuidholland in the thirteenth century, and Dutch colonists built similar mills in New York in the seventeenth century. All, however, were miniature operations.

The use of tidal power to generate large amounts of electricity has captured the popular fancy periodically over the last half century. In 1966, the French constructed the first commercial total power facility. The Saint-Malo plant on the Rance River, with a capacity of 240 megawatts, uses reversible bulb turbines to generate power both when the tide rises and when it falls.

Without droughts to plague it, tidal power has a seasonal advantage over hydroelectricity. Governed by the earth's rotation and the gravitational force of the moon, tides are comfortingly predictable. However, their periodicity causes formidable problems for those who would integrate tidal power into electrical utility systems. High tides occur about once every thirteen hours, and their peak power potential seldom coincides with peak power demand. The range between high and low tides changes on a semi-monthly cycle of forceful "springs" and weaker "neaps." At the Rance River plant, about four times as much power can be generated on the spring tides as on the neaps. Factoring this variable power source into an electrical system requires skillful planning and control.

Although its potential is limited to a small number of bays and estuaries with unusually high tides, tidal power has devoted followers around the world. The French are considering a 6,000-megawatt plant on the Bay of Mont-Saint-Michel. The Russians, having built a successful pilot plant at Kislaya Guba, are now exploring possible sites for several larger facilities. Canada and the United States are continuing their half-century study of the feasibility of exploiting some of the 30,000-megawatt potential of the Bay of Fundy. Potential tidal sites have been identified off the shores of twenty-three countries, including Australia, Argentina, China, and Korea, though several of these sites are at a considerable distance from current major energy markets.

Over the long term, tidal power probably constitutes one of the more environmentally sound energy sources. But siting limitations will severely restrict its importance, and tidal power can never provide more than 1 or 2 percent of the world's electrical capacity. While several proposed projects merit development, tidal power cannot, in the global scheme of things, be considered a major energy resource.

11. Plant Power: Biological Sources of Energy

GREEN PLANTS began collecting and storing sunshine more than two billion years ago. They photosynthesize an estimated one-tenth of one percent of all solar energy that strikes the earth. Somewhat more than half of this fraction is spent on plant metabolism; the remainder is stored in chemical bonds and can be put to work by human beings.

All fossil fuels were once biomass, and the prospect of dramatically shortening the time geological forces take to convert vegetation into oil, gas, and coal (roughly a third of a billion years) now intrigues many thoughtful persons. Dry cellulose has an average energy content of about 4 kilocalories per gram—60 percent as much as bituminous coal—and the hydrocarbons produced by certain plants contain more energy than coal does. Biomass can be transformed directly into substitutes for some of our most rapidly vanishing fuels.

Because green plants can be grown almost everywhere, they are not very susceptible to international political pressures. Unlike fossil fuels, botanical energy resources are renewable. In addition, biomass operations involve few of the environmental drawbacks associated with the large-scale use of coal and oil.

The ultimate magnitude of this energy resource has not been established. Measuring the earth's total photosynthetic capacity poses difficulties, and estimates vary considerably. Most experts peg the energy content of all annual biomass production at between fifteen and twenty times the amount human beings currently get from commercial energy sources, although other estimates range from ten to forty times.[1] Using all the vegetation that grows on earth each year as fuel is unthinkable.

But the energy that could reasonably be harvested from organic sources each year probably exceeds the energy content of all the fossil fuels currently consumed annually.

Two important caveats must be attached to this statement. The first qualification concerns conversion efficiency. Much of the energy bound in biomass will be lost during its conversion to useful fuels. These losses, however, need be no greater than those involved in converting coal into synthetic oil and gas. The second catch is geographical: the areas with the greatest biomass production are wet equatorial regions—not the temperate lands where fuel use is highest today. The full biological energy potential of the United States, calculated liberally, probably amounts to about one-fifth of current commercial energy use; in contrast, the potential in many tropical countries is much higher than their current fuel consumption levels. However, many equatorial nations will be hard pressed to secure the capital and to develop the technology needed to use their potential plant power.[2]

Organic fuels fall into two broad categories: waste from non-energy processes (such as food and paper production) and crops grown explicitly for their energy value. Since waste disposal is unavoidable and often costly, converting waste into fuels—the first option—is a sensible alternative to using valuable land for garbage dumps. However, the task of waste collection and disposal usually falls to those who cling to the bottom rungs of the economic and social ladder and, until recently, waste seldom attracted either the interest of the well-educated or the investment dollars of the well-heeled. But change is afoot, partly because solid waste is now often viewed as a source of abundant high-grade fuel that is close to major energy markets.

The wastes easiest to tap for fuels may be those that flow from food production. Bagasse, the residue from sugarcane, has long been used as fuel in most cane-growing regions. Cornstalks and spoiled grain are being eyed as potential sources of energy in the American Midwest. And India's brightest hope for bringing commercial energy to most of its 600,000 villages is pinned to a device that produces methane from excrement and that leaves fertilizer as a residue.

Wastes as Fuels

Agricultural residues—the inedible, unharvested portions of food crops—represent the largest potential source of energy from waste. But most plant residues are sparsely distributed, and some cannot be spared: they are needed to feed livestock, retard erosion, and enrich the soil. Yet, wisely used, field residues can guard the soil, provide animal fodder, *and* serve as a fuel source.

Agricultural energy demands are highly seasonal, and usage peaks do not always coincide with the periods during which residue-derived energy is most plentiful. In agricultural systems still largely dependent upon draft animals, this problem is minimized: silage and hay can easily be stored until needed. On mechanized farms, energy storage poses a somewhat more difficult problem.

Animal excrement is another potentially valuable source of energy. Much undigested energy remains bound in animal excrement; and cattle feedlots, chicken coops, and pigsties could easily become energy farms. Indeed, animal dung has been burned in some parts of the world for centuries; in the United States, buffalo chips once provided cooking fuel to frontiersmen on the treeless Great Plains. In India today, about 68 million tons of dry cow dung are burned as fuel each year, mostly in rural areas, although more than 90 percent of the potential heat and virtually all the nutrients in excrement are lost in inefficient burning.[3] Far more work could be obtained from dung if it were first digested to produce methane gas; moreover, all the nutrients originally in the dung could then be returned to the soil as fertilizer.[4]

In May, 1976, Calorific Recovery Anaerobic Process (CRAP), Inc., of Oklahoma City received Federal Power Commission authorization to provide the Natural Gas Pipeline Company annually with 820 million cubic feet of methane derived from feedlot wastes. Other similar proposals are being advanced. Although most commercial biogas plants planned in the United States are associated with giant feedlots, a more sensible long-term strategy might be to range-feed cattle as long as possible and then to fatten them up, a thousand at a time, on farms in the midwestern grain belt. Cow dung could power the farm and provide surplus methane, and the residue could be used as fertilizer. In addition, methane generation has been found to be economically attractive in most dairies—an important point, since more than half of all U.S. cows are used for milk production.[5]

Collectible crop residues and feedlot wastes in the United States contain 4.6 quadrillion Btu's (quads)—more energy than all the nation's farmers use.[6] Generating methane from such residues is often economical. However, developing a farm that is totally energy self-sufficient may require a broader goal than maximizing short-term food output.

Human sewage, too, contains a large store of energy. In some rural areas, particularly in China and India, ambitious programs to produce gaseous fuel from human and animal wastes are under way. Unfortunately, toxic industrial effluents are now mixed with human waste in many of the industrialized world's sewage systems, and these pollutants make clean energy recovery vastly more difficult. If these pollutants were kept separate, a large new energy source would become available.

The residues of the lumber and paper industries also contain usable energy. A study conducted for the Ford Foundation's Energy Policy Project found that if the U.S. paper industry were to adopt the most energy-efficient technologies now available and were to use its wood wastes as fuel, fossil fuel consumption could be reduced by a staggering 75 percent. The Weyerhaeuser Company recently announced a $75 million program to expand the use of wood waste as fuel for its paper mills; "We're getting out of oil and gas wherever we can," commented George Weyerhaeuser, the company's president. Sweden already obtains 7 percent of its total energy budget by exploiting wastes of its huge forest-products industry.

Eventually, most paper becomes urban trash. Ideally, much of it should instead be recycled—a process that would save trees, energy, and money. But unrecycled paper, along with rotten vegetables, cotton rags, and other organic garbage, contains energy that can be economically recaptured. Milan, Italy, runs its trolleys and electric buses partly on power produced from trash. Baltimore, Maryland, expects to heat much of its downtown business district soon with fuel obtained by distilling 1,000 tons of garbage a day.

American waste streams alone could, after conversion losses are subtracted, produce nearly five quads per year of methane and "char oil" —about 7 percent of the current U.S. energy budget. Decentralized agrarian societies could derive a far higher percentage of their commercial energy needs from agricultural, forest, and urban wastes.

Energy Crops

The second plant-energy option, the production of "energy crops," will probably be limited to marginal lands, since worldwide population pressures are already relentlessly pushing food producers onto lands ill-suited to conventional agriculture. Yet much potential energy cropland does exist in areas where food production cannot be sustained. Some prime agricultural land could also be employed during the off-season to grow energy crops. For example, winter rye (which has little forage value) could be planted in the American Midwest after the fall corn harvest and harvested for energy in the spring before maize is sowed.

Factors other than scarce land can limit biomass growth. The unavailability of nutrients and of an adequate water supply are two. Much marginal land is exceedingly dry, and lumber and paper industries will make large demands on areas wet enough to support trees. The energy costs of irrigating arid lands can be enormous, reducing the net energy output dramatically.

Yields from energy crops will reflect the amount of sunlight such crops receive, the acreage devoted to collecting energy, and the efficiency with which sunlight is captured, stored, harvested, transported, and put to work. Ultimately, they will also depend upon our ability to produce crops that do not sap the land's productivity and that can resist common diseases, pests, fire, and harsh weather.

The most familiar energy crop, of course, is firewood. A good fuel tree has a high annual yield when densely planted, resprouts from cut stumps (coppices), thrives with only short rotation periods, and is generally hardy. Favored species for fuel trees are eucalyptus, sycamore, and poplar—an intelligently planned tree plantation would probably grow a mixture of species.

Forests canopy about one-tenth of the planet's surface and represent about half the earth's captured biomass energy.[7] A century ago, the United States obtained three-fourths of its commercial energy from wood. In the industrialized world today, only a small number of the rural poor and a handful of self-styled rustics rely upon fuel wood. However, the case is emphatically different in the Third World. Thirty percent of India's energy and 96 percent of Tanzania's comes from wood.[8] In all, about half the trees cut down around the world are burned to cook food and to warm homes.

In many lands, unfortunately, human beings are propagating faster than trees. Although much attention has been paid to the population-food equation, scant notice has been given to the question of how the growing numbers will cook their food. As desperate people clear the land of mature trees and saplings alike, landscapes become barren, and, where watersheds are stripped, increasingly severe flooding occurs. In the parched wastelands of north central Africa and the fragile mountain environments of the Andes and the Himalayas, the worsening shortage of firewood is today's most pressing energy crisis.[9]

A variety of partial solutions have been suggested for the "firewood crisis." In southern Saudi Arabia, some tribes impose the same penalty for the unauthorized cutting of a tree as for the taking of a human life. China has embarked upon an ambitious reforestation program, and many other nations are following suit. Some forestry experts advocate substituting fast-growing trees for native varieties as a means of keeping up with demand.[10] However, the vulnerability of a forest of genetically similar trees to diseases and pests calls the application of such agricultural techniques to silviculture into question.

Improving the efficiency with which wood is used would also help alleviate the firewood shortage. In India, using firewood for cooking is typically less than 9 percent efficient. The widespread use of downdraft wood-burning stoves made of cast iron could, S. B. Richardson estimates, cut northern China's fuel requirements for heating and cooking by half.[11] Other efficient wood-burning devices can be made by local labor with local materials.

Wood can be put to more sophisticated uses than cooking and space heating. It can fuel boilers to produce electricity, industrial process steam, or both. The size of many prospective tree-harvesting operations (about 800 tons per day) is well tailored to many industrial energy needs. Decentralized co-generation using wood would also fit in well with current worldwide efforts to move major industries away from urban areas. In particular, the creation of forest "plantations" to produce fuel for large power plants at a cost comparable to that of coal has been recommended.[12] However, some researchers argue that the cost of transporting bulk biomass should lead us to think in terms of energy "farms" of a few thousand hectares or less.[13]

Trees are not the only energy crops worth considering. A number of other land and water crops have their advocates among bioconversion specialists. Land plants with potential as energy sources include sugarcane, cassava (manioc), and sunflowers, as well as some sorghums, kenaf, and forage grasses. Among the more intriguing plants under consideration are *Euphorbia lathrus* and *Euphorbia tirncalli*, shrubs whose sap contains an emulsion of hydrocarbons in water. While other plants also produce hydrocarbons directly, those produced by *Euphorbia* resemble the constituents in petroleum. Such plants might, Nobel laureate Melvin Calvin estimates, produce the equivalent of 10 to 50 barrels of oil per acre per year at a cost of $10 or less per barrel. Moreover, *Euphorbia* thrives on dry, marginal land.[14]

Several different crops could be cultivated simultaneously, a report by the Stanford Research Institute suggests, and side-by-side cropping could allow year-round harvesting in many parts of the world. Such mixed cropping would also increase ecological diversity, minimize soil depletion, and lower the vulnerability of energy crops to natural and human threats.[15]

Enthusiastic reports by NASA National Space Technology Laboratories have focused attention on the energy potential in water hyacinths. Thought to have originated in Brazil, the fast-growing water hyacinth now thrives in more than fifty countries; it flourishes in the Mississippi, Ganges, Zambezi, Congo, and Mekong rivers, as well as in remote irrigation canals and drainage ditches around the world. The government of Sudan is experimenting with the anaerobic digestion of thousands of tons of hyacinths mechanically harvested from the White Nile. However, a recent Batelle Laboratory report discounts the potential commercial importance of water hyacinths in the United States, in part because of their winter dormancy.[16]

Algae are another potential fuel. Some common types of this scummy, nonvascular plant have phenomenal growth rates. However, current harvesting techniques require large inputs of energy, the use of which lowers the net energy output of algae farming. Although solar drying would improve the energy balance, engineering breakthroughs are needed before impressive net energy yields can be obtained.

One of the more fascinating proposals for raising energy crops calls for the cultivation of giant seaweed in the ocean. As Dr. Howard Wilcox, manager of the Ocean Farm Project of the U.S. Naval Undersea Center in San Diego, points out, "Most of the earth's solar energy falls at sea, because the oceans cover some 71 percent of the surface area of the globe." The Ocean Farm Project, an effort to cultivate giant California kelp to capture some of this energy through photosynthesis, at present covers a quarter acre. But the experimental operation will, Wilcox hopes, eventually be replaced by an ocean farm 470 miles square. Such a sea field could theoretically produce as much natural gas as the United States currently consumes.[17]

Biomass Technologies

Biomass can be transformed into useful fuels in many ways, some of which were developed by the Germans during the petroleum shortages of World War II. Although one-third to two-thirds of the energy in biomass is lost in most conversion processes, the converted fuels can be used much more efficiently than raw biomass. The principal technologies now being explored are direct combustion, anaerobic digestion, pyrolysis, hydrolysis, hydrogasification, and hydrogenation.

In the industrialized world, organic energy is often recovered by burning urban refuse. To produce industrial process steam or electricity or both, several combustion technologies can be employed: waterwall incinerators, slagging incinerators, and incinerator turbines. Biomass can also be mixed with fossil fuels in conventional boilers, while fluidized-bed boilers can be used to burn such diverse substances as lumber-mill wastes, straw, corncobs, nutshells, and municipal wastes.

Since trash piles up menacingly in much of the urban world, cities can afford to pay a premium for energy-generating processes that reduce the volume of such waste. Urban trash lacks the consistency of coal, but its low sulfur content makes it an attractive energy source environmentally. Following the lead set by Paris and Copenhagen fifty years ago, several cities now mix garbage with other kinds of power-plant fuel to reduce their solid waste volume, to recover useful energy, and to lower the average sulfur content of their fuel. A $35 million plant in Saugus, Massachusetts, burns garbage from twelve towns, producing steam that is then sold to a nearby General Electric factory that hopes to save 73,000 gallons of fuel oil per day on its new fuel diet.

The next easiest method of energy recovery is anaerobic digestion —a fermenting process performed by a mixture of microorganisms in the absence of oxygen. In anaerobic digestion, acid-forming bacteria convert wastes into fatty acids, alcohols, and aldehydes; then methane-forming bacteria convert the acids to biogas. All biomass except wood can be anaerobically digested, and the process has been recommended for use in breaking down agricultural residues and urban refuse.[18] Anaerobic digestion takes place in a water slurry, and the process requires neither great quantities of energy nor exotic ingredients. Anaerobically digested, the dung from one cow will produce an average of 10 cubic feet of biogas per day—about enough to meet the daily cooking requirements of a typical Indian villager.

Many developing and some industrial nations are returning to this old technology, anaerobic digestion, for a new source of energy. Biogas generators convert cow dung, human excreta, and inedible agricultural residues into a mixture of methane and carbon dioxide that also contains traces of nitrogen, hydrogen, and hydrogen sulfide. Thirty thousand small biogas plants dot the Republic of Korea; and the People's Republic of China claims to have about two million biogas plants in operation.[19]

India has pioneered efforts to tailor biogas conversion to small-scale operations. After the OPEC price increases of 1973, annual gobar (the Hindi word for cow dung) gas plant sales shot up first to 6,560 and then to 13,000. In 1976, sales numbered 25,000. "We've reached takeoff," says H. R. Srinivasan, the program's director. "There's no stopping us now."

In addition to methane, other products can be derived from the biogasification of animal wastes and sewage. The residue of combustion is a rich fertilizer that retains all the original nutrients of the biomass and that also helps the soil retain water in dry periods. At Aurobindo Ashram in Pondicherry, India, wastes from cows, pigs, goats, and chickens will be gasified; the residue will be piped into ponds supporting algae, aquatic plants, and fish grown for use as animal fodder; and treated effluents from the ponds will be used to irrigate and fertilize vegetable gardens. Experience with biogas plants in "integrated farming systems" in Papua New Guinea suggests that the by-products of such controlled processes can be even more valuable than the methane.[20]

In developing countries, decentralized biological energy systems like that planned in Pondicherry could trigger positive social change. For small, remote villages with no prospects of getting electricity from central power plants, biogas can provide relatively inexpensive, high-grade energy and fertilizer. Ram Bux Singh, a prominent Indian developer and proponent of gobar gas plants, estimates that a small five-cow plant will repay its investment in just four years.[21] Larger plants serving whole villages are even more economically enticing. However, where capital is scarce, the initial investment is often difficult to obtain. In India, the Khadi and Village Industries Commission promotes gobar plant construction by granting subsidies and low-interest loans. The Commission underwrites one-fifth of the cost of individual plants and one-third of the cost of community plants. In the poorer areas, the Commission pays up to 100 percent of the cost of cooperative plants.

In efforts to hold down the cost of gobar plants and to conserve both scarce steel and cement in developing lands, researchers are producing new materials for use in digester construction. For example, a large cylindrical bag reinforced with nylon and equipped with a plastic inlet and outlet can be installed in a hole in the ground and weighted down in about one hour. The total cost can be as little as 15 percent of that of conventional digesters. Other experimental models are now being made out of natural rubber, mud bricks, bamboo pipes, and various indigenous hardwoods. In general, the ideal biogas plant for poor rural communities would be labor-intensive to build and operate and would be constructed of local materials.

The principal problem plaguing Third World biogas plants is temperature shifts, which can slow down or halt digestion. Low temperatures are particularly troublesome in Korea and China, where gas production slumps in winter when energy demands are highest. Possible remedies include improving insulation, burying future facilities to take advantage of subterranean heat, and erecting vinyl or glass green houses over the digesters to trap solar energy for heating. Alternatively, some of the gas produced in the digester could be used to heat the apparatus itself.

Alan Poole, a bioconversion specialist with the Institute for Energy Analysis at Oak Ridge, estimates that methane produced at the rate of 100 tons per day in a U.S. biogas plant would cost less than $4.00 per million Btu's, which approximates the expected cost of deriving commercial methane from coal.[22] In industrial countries, however, the recent trend has been away from anaerobic digestion. In 1963, this process was utilized in 70 percent of the U.S. wastewater treatment plants, but today it is being replaced—especially in smaller cities and towns—by processes that use more energy than they produce. The switch, which is now taking place at a capital cost in excess of $4 billion annually, was prompted largely by digester failures. Although poor design and operator error can both lead to pH imbalances or temperature fluctuations, the principal cause of unreliability appears to be the presence of inhibitory materials—especially heavy metals, synthetic detergents, and other industrial effluents.

These same industrial contaminants can also cause serious problems if the digested residues are used as fertilizer in agriculture. Some of these inhibitory substances can be separated routinely, but some will have to be cut off at the source and fed into a different treatment process if the excrement is to be anaerobically digested.

Anaerobic digestion produces a mixture of gases, only one of which —methane—is of value. For many purposes, the gas mixture can be used without cleansing. But even relatively pure methane is easy to obtain. Hydrogen sulfide can be removed from biogas by passing it over iron filings. Carbon dioxide can be scrubbed out with lime water (calcium hydroxide). Water vapor can be removed through absorption. The remaining methane has high energy content.

Biogas plants have few detractors, but some of their proponents fear that things are moving too fast and that large sums of money may be invested in inferior facilities when significant improvements may wait just around the corner. A recent report to the Economic Social Commission for Asia and the Pacific said of the Indian biogas program that "the cost should be drastically reduced, the digester temperature controlled during the winter months through the use of solar energy and the greenhouse effect, and the quality of the effluent improved," before huge amounts of scarce capital are sunk in biogas technology. To these misgivings must be added those of many in the Third World who are afraid that the benefits of biogas plants may fall exclusively or primarily to those who own cattle and land—accentuating the gap between property-owners and the true rural poor.[23]

To quell the fears of those with reservations about biogas development, most government programs stress community plants and cooperative facilities; and many countries are holding off on major commitments of resources to the current generation of digesters. But, whether small or large, sophisticated or crude, fully automated or labor-intensive, privately owned or public, biogas plants appear destined for an increasingly important role in the years ahead.

While hundreds of thousands of successful anaerobic digesters are already in operation, many other energy conversion technologies are also attracting increased interest. Hydrolysis, for example, can be used to obtain ethanol from plants and wastes with a high cellulose content at an apparent over-all conversion efficiency of about 25 percent. The cellulose is hydrolyzed into sugars, using either enzymes or chemicals; the sugar, in turn, is fermented by yeast into ethanol. Though most research on hydrolysis has thus far been small in scale, Australians have advanced proposals for producing prodigious quantities of ethanol using eucalyptus wood as the base and concentrated hydrochloric acid as the hydrolyzing agent. Ethanol so produced could substitute for a large share of Australia's rising oil imports.[24]

Pyrolysis is the destructive distillation of organic matter in the absence of oxygen. At temperatures above 500 degrees Centigrade, pyrolysis requires only atmospheric pressure to produce a mixture of gases, light oil, and a flaky char—the proportions of each being a function of operating conditions. In particular, this process recommends itself for use with woody biomass that cannot be digested anaerobically.

True pyrolysis is endothermic, requiring an external heat source. Many systems loosely termed "pyrolysis" are actually hybrids, employing combustion at some stage to produce heat. Three of the dozen or so systems now under development are far enough along to warrant comment. The Garrett "Flash Pyrolysis" process involves no combustion, but its end product (a corrosive and highly viscous oil) has a low energy content. The Monsanto "Langard" gas-pyrolysis process can be used to produce steam with an over-all efficiency of 54 percent. The Union Carbide "Purox" system, a high-temperature operation with a claimed efficiency of 64 percent, uses pure oxygen in its combustion stage and produces a low-Btu gas.[25]

Hydrogasification, a process in which a carbon source is treated with hydrogen to produce a high-Btu gas, has been well studied for use with coal. But further research is needed on its potential use with biomass, since, for example, the high moisture content of biomass may alter the reaction. Similarly, fluidized-bed techniques, which work well with coal, may require a more uniform size, shape, density, and chemical composition than biomass often provides. Experimental work on the application of fluidized-bed technologies to biomass fuels is now being conducted by the U.S. Bureau of Mines in Brucetown, Pennsylvania.

Hydrogenation, the chemical reduction of organic matter with carbon monoxide and steam to produce a heavy oil, requires pressures greater than 100 atmospheres. The U.S. Energy Research and Development Administration is paying for a $3.7 million pilot plant at Albany, Oregon; at the Albany plant, hydrogenation will be used to tap the energy in wood wastes, urban refuse, and agricultural residues.

Choice Fuels and Fuel Choices

The selection of energy systems will be partially dictated by the type of fuel desired: the ends will specify the means. In a sense, the development of biological energy sources is a conservative strategy, since the products resemble the fossil fuels that currently comprise most of the world's commercial energy use. Some fuels derived from green plants could be pumped through existing natural gas pipelines, and others could power existing automobiles. Nuclear power, in contrast, produces only electricity, and converting to an energy system that is mostly electric would entail major cultural changes and enormous capital expenditures.

Biomass processes can be designed to produce solids (wood and charcoal), liquids (oils and alcohols), gases (methane, hydrogen), or electricity. Charcoal, made through the destructive distillation of wood, has been used for at least ten thousand years. It has a higher energy content per unit of weight than does wood; its combustion temperature is hotter, and it burns more slowly. However, four tons of wood are required to produce one ton of charcoal, and this charcoal has the energy content of only two tons of wood. For many purposes—including firing boilers for electrical generation—the direct use of wood is preferable. Charcoal, on the other hand, is better suited to some specialized applications, such as steelmaking.

Methanol and ethanol are particularly useful biomass fuels. They are octane-rich, and they can be easily mixed with gasoline and used in existing internal-combustion engines. Both were commonly blended with gasoline, at up to 15 to 25 percent, respectively, in Europe between 1930 and 1950. Brazil recently embarked upon a $500 million program to dilute all gasoline by 20 percent with ethanol made from sugarcane and cassava. Meanwhile, several major U.S. corporations are showing keen interest in methanol. These alcohols could also fuel low-polluting external-combustion engines.[26]

The gaseous fuels produced from biomass can be burned directly to cook food or to provide industrial process heat. They can also be used to power pumps or generate electricity. Moreover, high-quality gases such as methane or hydrogen can be economically moved long distances via pipeline. A "synthesis gas" consisting of hydrogen and carbon monoxide was manufactured from coke in most U.S. towns at the turn of the century; known popularly as "town gas," it was piped to homes for lighting and cooking. A similar "local brew" might make sense today for areas rich in trees but poor in the biomass needed for anaerobic digestion. Synthesis gas can be further processed into methane, methanol, ammonia, or even gasoline.

The price in constant dollars for oil-based fuels declined during the 1950s and 1960s, partly because uses were found for more and more of the by-products of the refining process. Similarly, as the residues of biological energy processes find users, the production of fuels from biomass will grow more economically attractive.

Many biomass schemes reflect the assumption that energy crops can supply food as well as fuel. Even the plans to cultivate islands of deep-sea kelp include schemes for harvesting abalone in the kelp beds. Many energy crops, including water hyacinths, have proven palatable to cattle and other animals, once solar dryers have reduced moisture to appropriate levels.

More sophisticated by-product development has also been planned by students of chemurgy, the branch of applied chemistry concerned with the industrial use of organic raw materials. In the 1930s, George Washington Carver produced a multitude of industrial products from peanuts, while Percy Julian derived new chemicals from vegetable oils. And, for the record, the plastic trim on the 1936 Ford V–8 was made from soybeans.

Organic fuels can bear many different relationships to other products. Sometimes the fuels themselves are the by-product of efforts to produce food (e.g., sugar), natural fibers (e.g., paper), and lumber or wood chemicals (e.g., turpentine). Sometimes the residues of fuel-producing processes may be turned into plastics, synthetic fibers, detergents, lubricating oils, greases, and various chemicals.

Biological energy systems are free of the more frightening drawbacks associated with current energy sources. They will produce no bomb-grade materials or radioactive wastes. In equilibrium, biological energy sources will contribute no more carbon dioxide to the atmosphere than they will remove through photosynthesis; and switching to biomass conversion will reduce the cost of air pollution control, since the raw materials contain less sulfur and ash than many other fuels do. Indeed, some biological energy systems would have positive environmental impacts. Reforestation projects will control soil erosion, retard siltation of dams, and improve air quality. One type of biomass, water hyacinths, can control certain forms of water pollution, while others remove many air pollutants.

Without wise management, however, biological energy systems could engender major environmental menaces. The most elementary danger associated with biomass production is robbing the soil of its essential nutrients. If critical chemicals in the soil are not recycled, this "renewable" energy resource will produce barren wastelands.

Recycling nutrients can, alas, bring its own problems. First, if industrial wastes are included in the recycled material, toxic residues may build up in the soil. Some evidence suggests that certain contaminants —especially such heavy metals as cadmium and mercury—are taken up by some crops. Second, some disease-causing agents, especially viruses, may survive sewage treatment processes. Many of these potential infectants found in wastes can be controlled simply by aging the sludge before returning it to the soil. But during outbreaks of particularly virulent diseases, human excrement will have to be treated by other means, such as pasteurization, before being applied to agricultural lands.

Because of the relatively low efficiency with which plants capture sunlight, huge surfaces will be needed to grow large amounts of biomass. If biological energy farms significantly alter existing patterns of surface vegetation, the reflectivity and the water-absorption patterns of immense tracts of land could change. Moreover, new demands for gigantic tracts of land may eventually intrude upon public reserves, wetlands, and wilderness areas.

Ocean farming can go overboard too. The surface of the deep ocean is largely barren of plant nutrients, and large-scale kelp farming of the deep ocean might entail the use of wave-driven pumps to pull cold, nutrient-rich water from the depths up to the surface. A 100,000-acre farm might require the upwelling of as much as 2 billion tons of water a day, with unknown consequences for the marine environment. Deep waters also contain more inorganic carbon than surface waters do; upwelling such waters would entail the release of carbon dioxide into the atmosphere. (Ironically, a classic defense of biological energy systems has been that they would avoid the buildup of atmospheric CO_2 associated with the combustion of fossil fuels.) All these effects might be somewhat mitigated if ocean farms were located in cooler regions to the north and south, where the temperature difference between surface waters and deep waters is less.

If the quest for energy leads to the planting of genetically similar crops, the resulting monocultures will suffer from the threats that now plague high-yield food grains. Vulnerability to pests could necessitate widespread application of long-lived pesticides. An eternal evolutionary race would begin between plant breeders and blights, rots, and fungi. Moreover, biological energy systems are themselves vulnerable to external environmental impacts. A global cooling trend, for example, could significantly alter the growing season and the net amount of biomass an area could produce.

Using biomass conversion requires caution and respect for the unknown. If the expanded use of biological energy sources in equatorial countries resulted in the spread of harvesting technologies designed for use in the temperate zone, dire effects could follow. If the biomass fuels became items of world trade instead of instruments of energy independence, the sacking of Third World forests by multinational lumber and paper companies could be fatally accelerated.

The broad social effects of biological energy systems defy pat predictions. Biological energy systems could, for example, be designed to be labor-intensive and highly decentralized, but there is no guarantee that they will evolve this way of their own accord. Like all innovations, they must be carefully monitored; like all resources, they must be used to promote equity and not the narrow interests of the elite.

Photosynthetic fuels can contribute significantly to the world's commercial energy supply. Some of these solid, liquid, and gaseous fuels are rich in energy; and most can be easily stored and transported. Plant power can, without question, provide a large source of safe, low-polluting, relatively inexpensive energy. But all energy systems have certain intractable limits. For photosynthetic systems, these include the availability of sunlight and the narrowness of the radiation range within which photosynthesis can occur. Access to land, water, and nutrients will also set production boundaries. And, at a more profound level, we must ask how much of the total energy that drives the biosphere can be safely diverted to the support of a single species, *Homo sapiens*.

IV

Prospects and Consequences

12. Dawn of a New Era

W<small>E ARE</small> *not* running out of energy. However, we *are* running out of cheap oil and gas. We are running out of money to pay for doubling and redoubling an already vast energy supply system. We are running out of political willingness to accept the social costs of continued rapid energy expansion. We are running out of the environmental capacity needed to handle the waste generated in energy production. And we are running out of time to adjust to these new realities.

For two decades, we have pursued a chimerical dream of safe, cheap nuclear energy. That dream has nearly vanished. Nuclear fission now appears to be inextricably bound to weapons proliferation and to a broad range of other intractable problems. Every week new evidence buttressing the case against nuclear power is uncovered; every week worldwide opposition to nuclear power grows stronger. Nuclear fission now appears unlikely ever to contribute a large fraction of the world's energy budget.

Humankind is consequently no closer today than it was two decades ago to finding a replacement for oil. Yet the rhetoric that public officials in the world's capitals lavish upon the energy "crisis" is not being translated into action. Most energy policy is still framed as though it were addressing a problem that our grandchildren will inherit. But the energy crisis is *our* crisis. Oil and natural gas are our principal means of bridging today and tomorrow, and we are burning our bridges.

Twenty years ago, humankind had some flexibility; today the options are more constrained. All our possible choices have long lead times. All new energy sources will require new factories to produce new equipment and large numbers of workers with new skills. Energy conservation programs will similarly require decades to implement fully, as existing inventories of energy-using devices are slowly replaced. Inefficient buildings constructed today will still be wasting energy fifty years from now; oversized cars sold today will still be wasting fuel ten years down the road.

If the energy crisis has no "quick fix," neither is there any long-term *deus ex machina*. Great progress has been made on coal conversion technologies in recent years, but environmental and resource constraints necessarily limit coal to a transitional role. Coal can and should be substituted for oil and gas in many instances, but coal cannot replace the 75 percent of all commercial energy these fuels now provide.

Nuclear fusion, if feasible at all, would be expensive, incredibly complex, and highly centralized. For technical reasons, the first generation of fusion reactors would probably consist of fusion-fission hybrids designed to breed plutonium. Such devices would lead the world into an unconscionable "plutonium economy" and will therefore be vigorously fought by a formidable array of opponents. While "pure" fusion deserves continued research support, it holds no immediate potential, and even over the long term there is no assurance that it will become a commercially viable source of power.

Although no easy answers exist, some solutions clearly outshine others. Of the supply technologies in hand today, solar, wind, water, and biomass sources appear most attractive. And for years to come, the world's greatest opportunities will lie in energy conservation.

Priorities for a Post-Petroleum World

The energy crisis demands rapid decisions, but policies must nevertheless be formulated with an eye to their long-term implications. In making each of hundreds of discrete decisions, we would be well advised to apply a few basic criteria. Thrift, renewability, decentralization, simplicity, and safety should be the touchstones. Using these, we might judge whether a given action will move us closer to, or further from, the type of energy system we ultimately seek.

Both rich industrial countries and poor agrarian ones can cull far more benefits in the immediate future from investments in increased efficiency than from investments in new energy sources. In fact, because they are unable to afford to make the necessary initial investments that conservation sometimes requires, the poor frequently waste a higher fraction of the energy they use than do the well-to-do. By eliminating waste and by matching energy sources carefully with appropriate uses, people can wring far more work from every unit of energy than is now the case. A sensible energy strategy will help accomplish this sensible goal.

Energy is a means, not an end. Its worth derives entirely from its capacity to perform work. No one wants a kilowatt-hour; the object is to light a room. No one wants a gallon of gasoline; the object is to travel from one place to another. If our objectives can be met using a half, or even a quarter, as much energy as we now use, no benefit is lost.

Investments in conservation must mesh with plans for a rapid switch from fossil fuels to sustainable energy sources. An intelligent strategy will lead to dependence upon energy derived solely from perpetually reliable sources. Solar technologies alone can provide us with as much energy as can be safely employed on our fragile planet.

In establishing priorities for the post-petroleum period, foremost attention should be given to basic human needs—to food, shelter, clothing, health care, and education. Fortunately, such needs either require comparatively little energy or have energy requirements that can be met with renewable energy sources. Indeed, for most of history *Homo sapiens* has been entirely dependent upon renewable energy sources, and could not have survived if renewable sources had not met the most basic needs.

The industrial world, powered mostly by renewable energy sources a mere hundred years ago, now runs almost entirely on fossil fuels. The agrarian nations still obtain more than two-thirds of their fuel from sustainable sources—mostly firewood and forage for draft animals. These two worlds consequently face different problems, and may honor different priorities during the coming transition.

In the Third World, enormous strides can be made with relatively modest investments if those investments are made wisely. For example, 2 percent of the world military budget for just one year could provide every rural Third World family with an efficient stove—doubling overnight the amount of useful work obtained from fuel wood, and reducing the pressure on the world's forests accordingly. If, in addition, armies were mobilized in major tree-planting campaigns, the firewood crisis could eventually be alleviated.

In the industrial world, the situation is arguably more precarious, and dramatic steps are in order. However, such steps are not being taken. For example, a responsible energy policy reflecting the urgency of the necessary transition would require that *all* new automobiles average at least 35 miles per gallon within three years, and that the transition to non-petroleum vehicles be well under way within a decade. If the energy transition were proceeding on a reasonable timetable, tens of millions of solar water heaters would be produced annually; current production, by contrast, is in the thousands. While the generation of electricity from high-temperature industrial steam is the cheapest and most attractive new power source in many countries, institutional factors have caused this technology to be slighted all over the world.

It is virtually impossible to develop a list of global energy priorities. Each country must pursue those options most compatible with its conditions and its aspirations. But in general, conservation investments will prove more immediately productive than new source development, and genuine necessities, such as food, must always take precedence over frivolous trimmings.

Suitable Energy Technologies

Historically, many important inventions have consisted of no more than ingenious new applications of existing knowledge. In recent decades, however, large teams of specialists wielding complex and expensive research tools have been increasingly rubbing against the boundaries of knowledge. Nowhere is this phenomenon more clear than in the industrial world's response to the energy crisis. Research is currently focused on the liquid metal fast breeder reactor, with fusion reactors and coal conversion technologies vying for the remaining funds. Sources that don't cost billions of dollars to develop seem almost unworthy of serious consideration. The "hard" technologies obtain the most funds, attract the brightest researchers, kindle the greatest public interest, and accrue the most glamour. They do not, however, necessarily represent the wisest choices. Nuclear fusion research may well yield a Nobel Prize someday; no plausible line of research on biogas plants seems likely to win a trip to Stockholm. Nevertheless, biogas plants will almost certainly provide more energy to those who need it most than fusion reactors ever will.

Energy funding continues to be apportioned as though big were beautiful, and the reasons for this are understandable. "Those in power always want big accomplishments—scientific breakthroughs and politically visible facilities," explains M. C. Gupta, director of the Thermodynamic Laboratory at the Indian Institute of Technology. "But those things aren't what India needs most. The needs of our neediest can only be met by small, inexpensive devices that use indigenous materials and are easily maintained."

Even research on direct and indirect solar sources will not necessarily produce devices that meet the diverse needs of the world's peoples. Every technology embodies the values and conditions of the society it was designed to serve. Most significant research on sustainable energy sources has been performed in industrialized countries. Technological advances have therefore reflected the needs of societies with temperate climates, high per capita incomes, abundant material resources, sophisticated technical infrastructures, expensive labor, good communication and transportation systems, and well-trained maintenance personnel. Such societies are wired for electricity—indeed, two-thirds of the U.S. solar energy research budget is devoted to the generation of electricity.

Clearly, some of the findings of this research are not easily or wisely transferred to societies with tropical climates, low per capita incomes, few material resources, stunted technical infrastructures, cheap labor, poor communications, and only fledgling maintenance forces. Most people in the world do not have electrical outlets or anything to plug into them. What they need are cheap solar cookers, inexpensive irrigation pumps, simple crop dryers, small solar furnaces to fire bricks, and other basic tools.

With the traps of technology transfer in mind, some argue that a major solar research and development effort on the part of the industrialized world is irrelevant to the true needs of the poorer countries. This argument contains a kernel of truth in a husk of misunderstanding. Countries can choose to learn from each other's experience, but each country must view borrowed knowledge through the lens of its own unique culture, resources, geography, and institutions. The United States and China can trade knowledge to good purpose, but little of what they trade can be transplanted intact.

The differences between such industrialized lands as Japan and France merit note, but the differences between two Third World countries may be more striking than the similarities. Surinam (with an annual per capita income of $810) has energy problems and potential solutions to those problems that bear little resemblance to those of Rwanda (with an annual per capita income of about $60). And national wealth is not the only feature in an energy profile. The tasks for which energy is needed vary from country to country. In some, the most pressing need may be for pumps to bring water from a deep water table to the parched surface. In lands with more abundant water supplies, cooking fuel may be in desperately short supply. The availability of sustainable resources may also differ. One region may have ample hydropower potential, another strong winds, and a third profuse direct sunlight. Successful technology transfers require a keen sensitivity to such differences.

Some disillusioned solar researchers in both industrialized and agrarian countries contend that the major impediment to solar development has been neither technical (the devices work) nor economic (many simple devices can be cheaply made). Instead, they claim, the problems have social and cultural roots. Many Third World leaders did not want to settle for "second-rate" renewable energy sources while the industrial world flourished on oil and nuclear power. Often, officials who found themselves in charge of new technologies, such as windmills, were unable to find technicians who could maintain and repair them. Occasionally, people who were given solar equipment refused to use it because the rigid time requirements of solar technology disrupted their daily routines or because the direct use of sunlight defied their cultural traditions.

Many of these attitudinal impediments may now be vanishing as the global south begins developing its own research and development capacity. The indigenous technologies born of the new capability may prove to be more compatible with Third World needs than borrowed machines and methods. Brazil's large methanol program, India's gobar gas plants, and the Middle East's growing fascination with solar electric technologies can all be read as signs of an interest in renewable energy resources that bodes well for the future. At the same time, the Third World, stunned by a simultaneous shortage of firewood and petroleum, may be more willing than it was a few years ago to adopt solar solutions.

In much of the global north as well, solar technologies are being embraced as important future options. In Japan, the Soviet Union, France, and the United States, renewable resources are increasingly being viewed as major components of future energy planning. Some of the innovative research in these countries could well be of global significance.

Energy and International Equity

The world's most lamentable social problem is doubtless the enduring hypocrisy of poverty. Although well-publicized conferences periodically issue calls for "development decades," foreign aid "targets," and other high-sounding programs, the gap in the absolute income between rich and poor countries grows steadily wider.

Decisions on energy sources can dramatically affect the international distribution of wealth. High-priced oil, for example, has brought a flood of dollars—mostly from the rich industrial countries—to what had previously been some of the world's poorest lands. The rest of the Third World, although itself hard hit by rising oil prices, has rather steadfastly maintained its solidarity with the oil exporting countries; rising prices for raw materials are viewed as crucial components of a far-reaching new economic order, and oil is currently the world's most important raw material. Other countries that export natural resources hope that OPEC's successful price hikes will blaze a trail they can follow.

Although the new economic order is generally defined in terms of commodity prices and monetary reforms, its success may hinge on the choice of a post-petroleum energy source. Whereas complex technologies would divert a major stream of scarce capital to the industrial world, the development of safe sustainable sources could cause investment dollars to flow in the other direction. Direct and indirect solar sources thus appear to hold a double economic promise for the Third World.

Investment funds tend to become available where energy is available. Industries compete vigorously for the right to build plants in the Middle East, less to penetrate the region's small markets than to be assured of a supply of fuel. As renewable sources attract more adherents, hard currencies can be expected to flow to the world's richest sources of sunlight, wind, water, and biomass, and most of these are located in the Third World.

Foreign investments can hold pitfalls for the unwary. Ghana, for example, was able to attract British and American financing for the Volta River dam only after Kaiser Aluminum entered into a long-term contract to buy 310 of the 540 megawatts produced by the dam. This arrangement permitted Ghana to finance the development of an important renewable energy source, but the costs to Ghana were steep: relocating 80,000 people dislocated by the reservoir and battling the rampant parasite disease schistosomiasis. Kaiser's aluminum refinery uses more than half the electricity produced at the dam, and the benefits to Ghana are few: the plant produces little employment, and the refined aluminum is shipped out as ingots, not as manufactured goods. Ghana, despite its large bauxite reserves, does not even derive a secondary benefit as a raw material vendor, since Kaiser imports all its ore from mines in the West Indies. Far from being the centerpiece of a comprehensive national development strategy, Ghana's dam is little more than a means of harnessing African water power to serve the needs of the industrial world.

If resource exporting countries are to enter fully into a new economic order, they must be able to process much of the material they produce, tapping locally available flows of energy. In an era of diffuse energy resources, the enormous use of energy that now characterizes the industrial world would be spread out over the entire globe. Instead of shipping ore to Europe for refining, the producing country would ship refined metal. Containing "embodied" energy derived from natural sources, the refined metal is worth much more than ore, so the exporting country would achieve a more favorable balance of trade. As an industrial infrastructure takes shape, the exporting country would also be able to produce and sell more manufactured products.

Poverty is, of course, a matter of people as much as of countries. Almost all poor countries have some rich people, and all rich countries have poor people. Increases in national income do not necessarily mean that the new wealth will be shared. In some oil producing countries, rising revenues have left the rich richer and the poor untouched.

If vigorous conservation is to lead eventually to an energy ceiling, population growth must be constrained as energy is equalized. The alternative is to divide a constant amount of energy among an ever-increasing pool of people. Population stabilization is imperative both in the industrial world, where non-renewable fuel consumption per person is twenty to thirty times higher than in the Third World, and in the Third World, where burgeoning population growth is outstripping traditional energy sources such as firewood. Like energy itself, population is a global problem, and it requires a worldwide solution.

The development of renewable energy sources cannot itself abolish poverty—only widespread social and political change can. But decentralized sources of energy are compatible with a development strategy that grows from the bottom up, rather than one that merely permits a few benefits to trickle down to the masses from the elite in control of centralized high technologies. The use of appropriate energy sources will facilitate a more equitable distribution of wealth and power both within and among nations, by transferring control from distant corporations and bureaucracies to more responsive local units.

Energy and the Human Prospect

For twenty years, the world has pursued a dead-end path. This energy route cannot be changed without fundamentally altering society. Some alternatives are better than others because the changes they dictate are relatively attractive, but there is no way of avoiding some form of pervasive change. If, for example, the world were to opt for harmonious, small-scale, decentralized, renewable energy technologies, few aspects of modern life would go unaffected.

Farms would begin to supply large fractions of their own energy through wind power, solar heaters, and technologies for harnessing the energy in agricultural wastes. Such self-sufficient farms would tend to be smaller and to provide more employment than those that prevailed in the oil era. Food storage and preparation would slowly be shifted to solar-powered technologies. Meat consumption in the industrial world would drop and the food processing industry would become more energy-efficient and less pervasive in its impact on diets.

In the new energy era, transportation would be weaned from its petroleum base even as improved communications and intelligent city planning began to eliminate pointless travel. Energy efficiency and load factors would become important criteria in evaluating transport modes, and would be reflected in the costs of travel. Bicycles would begin to account for an important fraction of commuter traffic as well as of other short trips. And freight transport would be transferred wherever possible to more energy-efficient modes, especially trains and ships.

If we were to opt for the best renewable energy technologies, buildings could be engineered to take full advantage of their environments. More and more of the energy needed for heating and cooling would be derived directly from the sun. Using low-cost photovoltaics that convert sunlight directly into electricity, many buildings could eventually become energy self-sufficient. New jobs and professions would develop around the effort to exploit sunlight, and courts would be forced to consider the "right" of building owners not to have their sunshine blocked by neighboring structures.

While industry would doubtless turn to coal for much of its energy during the transition period, eventually it would also draw its primary energy from natural flows. Thus, energy availability would play an important role in determining the locations of future factories. The sunshine-rich nations of the Third World, where raw materials and renewable energy sources are most plentiful, could become new centers of economic productivity. The across-the-board substitution of cheap fuel for human labor would be halted. Recycled metals, fibers, and other materials would become principal sources of raw materials. Seen as energy repositories, manufactured products would necessarily become more durable and would be designed to be easily repaired and recycled.

Using small, decentralized, and safe technologies makes sense from a systems management point of view. Small units could be added incrementally if rising demands required them, and they would be much easier than large new facilities to integrate smoothly into an energy system. Small, simple sources could be installed in a matter of weeks or months; large, complex facilities often require years and even decades to erect. If gigantic power plants were displaced by thousands of smaller units dispersed near the points of end use, economies of size would become relatively less important vis-à-vis economies of mass production.

Technology would again concern itself with simplicity and elegance, and vast systems would become extinct as more appropriately scaled facilities evolved.

To decentralize power sources is in a sense to act upon the principle of "safety in numbers." When large amounts of power are produced at individual facilities or clusters of plants, the continued operation of these plants become crucial to society. Where energy production is centralized, those seeking to coerce or simply to disrupt the community can easily acquire considerable leverage: for example, a leader of the British electrical workers recently noted that "the miners brought the country to its knees in eight weeks," but that his co-workers "could do it in eight minutes." Disruption need not be intentional, either. Human error or natural phenomena can easily upset fragile energy networks that serve wide areas, while use of diverse decentralized sources could practically eliminate such problems.

The societies that will develop around efficient, renewable, decentralized, simple, safe energy sources cannot be fully visualized from our present vantage point. Indeed, one of the most attractive promises of such sources is a far greater flexibility in social design than is afforded by their alternatives. Though energy sources may not dictate the shape of society, they do limit its range of possibilities, and diverse, dispersed energy sources are more compatible than centralized technologies with social equity, freedom, and political participation.

Societies based upon natural flows of energy will have to wrestle with the concept of limits. Endless and mindless growth is not possible for nations living on energy income instead of capital. Such societies will need public policies and ethics that disparage rather than whet the appetite for frivolous consumption. Materialism, which gives sanction to what Voltaire saw as humanity's perpetual enemies—poverty, vice, and boredom—will need to be replaced by a new source of social vitality that is less corrosive to the human spirit and less destructive to the collective environment.

The attractions of sunlight, wind, running water, and green plants as energy sources are self-evident. They are especially appealing in their stark contrast to a world of nuclear garrison states. Scarce resources would be conserved, environmental quality would be maintained, and employment would be spurred. Decentralized facilities would lead to a more local autonomy and control. Social and financial equity would be increased, within and among nations.

Had industrial civilization been built upon such forms of energy "income" instead of on the energy stored in fossil fuels, any proposal to convert to coal or uranium for the world's future energy would doubtless be viewed with incredulous horror. The current prospect, however, is the reverse—a shift from trouble-ridden sources to more attractive ones. Of the possible worlds we might choose to build, an efficient solar-powered one appears most inviting.

Notes

Chapter 1. Introduction: Twilight of an Era

1. Stephen H. Schneider with Lynne E. Meisrow, *The Genesis Strategy: Climate and Global Survival* (New York: Plenum Press, 1976). Bert Bolin, *Energy and Climate* (Stockholm: Secretariat for Future Studies, 1975).

2. Paul E. Damon and Steven M. Kunen, "Global Cooling?," *Science*, vol. 193, no. 4252 (Aug. 6, 1976).

3. The most comprehensive recent study of oceanic oil is U.S. National Academy of Sciences, *Petroleum in the Marine Environment* (Washington, D.C., 1975). The special problems of offshore oil exploitation are explored through a case study in Scotland in Pamela L. Baldwin and Malcom E. Baldwin, *Onshore Planning for Offshore Oil* (Washington, D.C.: Conservation Foundation, 1975). Some of the international political implications of oil spills are explored in Sheldon Novick, "Ducking Liability at Sea," *Environment*, Jan.-Feb., 1977. Probably the best over-all introduction to the problems of oil supertankers is Noël Mostert's *Supership* (New York: Knopf, 1974).

4. M. Blumer, "Oil Contamination and the Living Resources of the Sea," presented to the FAO Technical Conference on Marine Pollution and Its Effects on Living Resources and Fishing (Rome, Dec. 9–18, 1970). The pollution of the world's waterways by oil is, of course, only an illustrative example of the environmental burdens of increasing energy use. For a comprehensive overview of the broad issue, see Paul R. Ehrlich, Anne H. Ehrlich, and John Holdren, *Population, Resources, Environment* (San Francisco: W. H. Freeman, 1977).

5. Richard S. Claassen, "Materials for Advanced Energy Technologies," *Science*, Feb. 20, 1976.

6. U.S. Water Resources Council, *Water for Energy Self-Sufficiency* (Washington, D.C., 1974).

7. Barry Bosworth, James S. Duesenberry, Andrew S. Carron, *Capital Needs in the Seventies* (Washington, D.C.: Brookings Institution, 1975). Peter L. Bernstein, "Capital Shortage: Cyclical or Secular?," *Challenge*, vol. 8, no. 5 (Nov.-Dec., 1975). Henry C. Wallich, "Is There a Capital Shortage?," *Challenge* vol. 8, no. 4 (Sept.-Oct., 1975). Arthur D. Little, Inc., *Capital Needs and Federal Policy Choices in the Energy Industries*, submitted to the Federal Energy Administration, Aug. 16, 1974. Jerome E. Hass, Edward J. Mitchell, Bernell K. Stone, with David H. Downes, *Financing the Energy Industry*, a report to the Energy Policy Project of the Ford Foundation (Cambridge, Mass.: Ballinger, 1974).

8. M. Carasso et al., *The Energy Supply Planning Model*, PB 245–382 and PB 245–383 (Springfield, Va.: NTIS, Aug., 1975).

9. Carol J. Loomis, "For the Utilities, It's a Fight for Survival," *Fortune*, March, 1975.

10. John E. Gray, *Financing Free World Energy Supply and Use* (Washington, D.C.: Atlantic Council, Feb. 11, 1975).

11. David L. Ostendorf with Joan E. Gibson, *Illinois Land: The Emerging Conflict over the Use of Land for Agricultural Production and Coal Development* (Carterville, Ill.: Illinois South Project, 1976). John C. Doyle, Jr., *Strip Mining in the Corn Belt: The Destruction of High Capability Agricultural Land for Strip-Minable Coal in Illinois* (Washington, D.C.: Environmental Policy Institute, June, 1976).

12. Department of Economic and Social Affairs, *World Energy Supplies: 1950–1974* (New York: United Nations, 1976).

13. Robert C. Axtmann, "Environmental Impact of a Geothermal Power Plant," *Science*, vol. 187, no. 4179 (March 7, 1975).

14. An overview of the major components of the U.S. fusion program can be obtained from the Energy Research and Development Administration, *Fusion Power by Magnetic Confinement Program Plan*, vols. I, II, III, IV (Washington, D.C., July, 1976). For an excellent survey of the technical problems faced by fusion written from an optimistic viewpoint, see David J. Rose and Michael Feirtag, "The Prospect for Fusion," *Technology Review*, Dec., 1976. For a more skeptical appraisal, see the three-part series by William Metz, "Fusion Power: What Is the Program Buying the Country?," *Science*, June 25, 1976; "Fusion Research: Detailed Reactor Studies Identify More Problems," *Science*, July 2, 1976; "Fusion Research: New Interest in Fusion-Assisted Breeders," *Science*, July 23, 1976.

Chapter 2. The Future of Fossil Fuels

1. Joel Darmstadter, with Perry Teitelbaum and Jaroslav Pollach, *Energy in the World Economy: A Statistical Review of Trends in Output, Trade, and Consumption since 1925* (Baltimore: Johns Hopkins Press for Resources for the Future, 1971). Department of Economic and Social Affairs, *World Energy Supplies: 1950–1974*, Statistical Papers, ser. J, no. 19 (New York: United Nations, 1976).

2. Robert Engler, *The Politics of Oil: A Study of Private Power and Democratic Directions* (Chicago: University of Chicago Press, 1961).

3. M. King Hubbert, "Nuclear Energy and the Fossil Fuels," *Drilling and Production Practice* (American Petroleum Institute, 1956; reprinted by Shell Oil Company).

4. An unusually lucid discussion of resource terminology can be found in V. E. McKelvey, "Mineral Resource Estimates and Public Policy," *American Scientist*, Jan.–Feb., 1972. A different perspective is presented by D. C. Ion in his recent, comprehensive volume, *Availability of World Energy Resources* (London: Graham & Trotman, 1976).

5. Congressional Research Service, *Secondary and Tertiary Recovery of Oil*, a report to the Subcommittee on Energy of the U.S. House of Representatives Committee on Science and Astronautics (Washington, D.C.: Government Printing Office, 1974).

6. Betty M. Miller, Harry L. Thomsen, Gordon L. Dalton, Anny B. Coury, Thomas and Katharine L. Varnes, "Geological Estimates of Undiscovered Recoverable Oil and Gas Resources in the United States," U.S. Geological Survey Circular 725. Comrate, *Mineral Resources and the Environment*, report prepared by the Committee on Mineral Resources and the Environment, Commission on Natural Resources, National Research Council, National Academy of Sciences (Washington, D.C., Feb., 1975).

7. The fuel resource figures used in this chapter were derived from D. C. Ion, op. cit.; *Survey of Energy Resources*, by World Energy Conference, Detroit, 1974; and M. King Hubbert, "Energy Resources," in *Resources and Man*, for the National Academy of Sciences (San Francisco: W. H. Freeman, 1969).

8. Bernard Grossling of the U.S. Geological Survey believes that Latin American resources may be more than twice this size. *Latin America's Petroleum Prospects in the Energy Crisis*, U.S. Geological Survey Bulletin 1411, 1975.

9. In my conversations with the Saudi Arabian foreign minister and petroleum minister, these themes came up repeatedly.

10. T. D. Adams and M. A. Kirby, "Estimate of World Gas Reserves," IX W.P.C. Preprint, P.D. 6(1), 1975, cited in D. C. Ion, op. cit.

11. Environmental Studies Board of the U.S. National Academy of Sciences, *Rehabilitation Potential of Western Coal Lands*, report to the Energy Policy Project of the Ford Foundation (Cambridge, Mass.: Ballinger, 1974). Robert Stefanko, R. V. Ramani, and Michael R. Ferko, *An Analysis of Strip Mining Methods and Equipment Selection*, report to the U.S. Office of Coal Research under Contract No. 14-01-0001-390, May 29, 1973. E. A. Nephew and R. L. Spore, *Costs of Coal Surface Mining and Reclamation in Appalachia*, Oak Ridge National Laboratory Report No. ORNL-NSF-EP-86, Jan., 1976. The political dimensions of this issue are explored in Marc Karnis Landy, *The Politics of Environmental Reform: Controlling Kentucky Strip Mining* (Washington, D.C.: Resources for the Future, 1976).

12. R. R. Ruch, H. J. Gluskoter, and N. F. Shimp, *Occurrence and Distribution of Potentially Volatile Trace Elements in Coal*, Environmental Geology Notes, No. 72, Aug., 1974, Illinois State Geological Survey. D. F. S. Natusch et al., "Toxic Trace Elements: Preferential Concentration in Respirable Particles," *Science*, vol. 183 (Jan. 18, 1974).

13. Synfuels Interagency Task Force, *Draft Environmental Impact Statement on Synthetic Fuels Commercialization Program* (Washington, D.C.: Government Printing Office, Dec., 1975). U.S. Federal Energy Administration, *Draft Environmental Impact Statement on a Coal Conversion Program* (Jan., 1975).

14. Earl Cook, "Limits to Exploitation of Nonrenewable Resources," *Science*, vol. 191, no. 4228 (Feb. 20, 1976).

15. Arthur M. Squires, "Chemicals from Coal," *Science*, vol. 191, no. 4228 (Feb., 1976).

16. P. Chapman, G. Leach, and M. Slesser, "The Energy Costs of Fuels," *Energy Policy*, Sept., 1974.

17. Arthur M. Squires used the Spanish gold metaphor in "Coal: A Past and Future King," *Ambio*, vol. 3, no. 1 (1974).

Chapter 3. Nuclear Power: The Fifth Horseman

1. Atomic Industrial Forum, "Nuclear Power Plants outside the United States," June 10, 1977. *World Environment Report*, June 9, 1975 (special nuclear issue). *Market Survey for Nuclear Power in Developing Countries* (Vienna: International Atomic Energy Agency, 1974). A number of studies have shown that, owing to the large energy investments needed to build and fuel a nuclear power plant before it can begin operations, such a rate of nuclear growth would result in a net energy *drain* for the next twenty-five years. See, for example, John Price, "Dynamic Energy Analysis and Nuclear Power," in Amory B. Lovins and John H. Price, *Non-Nuclear Futures* (Cambridge, Mass.: Ballinger, 1975).

2. An overview of the Swedish position can be found in Lennart Daleus, "A Moratorium in Name Only," *Bulletin of the Atomic Scientists*, Oct., 1975; U.S. figures were obtained from Atomic Industrial Forum, op. cit., and *Weekly Energy Report*, Jan. 5, 1975; the *Japan Times* has given much coverage to nuclear issues, and information on the *Mutsu* was obtained from the Jan. 16, 17, 19, 22, and Feb. 11, 17, 1976 issues; most other national data are from various issues of *Not Man Apart*, a publication of the U.S. branch of Friends of the Earth, and from the New York *Times* and Washington *Post*.

3. In 1975, nuclear exports amounted to $3.6 billion, two-thirds of which were U.S. sales (*Economist*, Dec. 6, 1975). William Casey, president of the Export-Import Bank, predicts that "within the next three years nuclear technology will become the U.S. economy's biggest export item." By far the most comprehensive analysis of nuclear exports is in Richard J. Barber Associates, *L.D.C. Nuclear Power Prospects, 1975–1990*, ERDA-52 (Springfield, Va.: NTIS, 1975). For insight into the attitude of the American business community on this issue, see Tom Alexander, "Our Costly Losing Battle against Nuclear Proliferation," *Fortune*, Dec., 1975.

4. The most common reactors—light water reactors—require "enriched" uranium, fuel that is 3 to 4 percent U 235. Yet natural uranium contains only 0.7 percent U 235.

Further, this isotope is chemically identical with the far more common form, U 238, and cannot be separated by simple chemical reactions. Elaborate physical enrichment processes that can distinguish between atoms on the basis of weight are needed to separate U 238 atoms, which constitute 99.3 percent of all natural uranium, from the infinitesimally lighter U 235 atoms.

For the past two decades, the United States has dominated the world market for enriched uranium, with production from three large gaseous diffusion plants at Oak Ridge, Tennessee; Paducah, Kentucky; and Portsmouth, Ohio. But now numerous other countries are experimenting with several new enrichment technologies. Four general enrichment processes are in differing stages of development: gaseous diffusion, centrifuge, nozzle, and laser.

In centrifuge enrichment, uranium hexafluoride gas is fed into a spinning centrifuge. Here U 238 is spun toward the walls while the lighter U 235 passes into an upper chamber. The principal advantage of centrifuge enrichment is an energy requirement equal to about 10 percent that required by the gaseous diffusion process.

Centrifuge enrichment has been successfully developed to the pilot-plant stage, and Urenco, Ltd., a British–Dutch–West German collaboration, is building one small facility at Capenhurst, England, and another at Almelo, Holland. In the United States, the federal government is constructing a slightly larger centrifuge facility at Oak Ridge, using a different technology, and three U.S. corporations are also seeking to enter the field.

In nozzle enrichment, a jet of uranium hexafluoride gas is squirted into a low-pressure tank. The heavier U 238 tends to flow straight to a "paring" tube on the other side of the tank, while the lighter U 235 tends to drift to the side.

Nozzle enrichment was developed in West Germany in the mid-1950s. Nozzle enrichment facilities should be comparatively easy to engineer, although their total energy requirements will be rather high. South Africa has done much of the commercial development of nozzle enrichment, and West Germany has contracted to build such a plant in Brazil.

Laser enrichment, the least developed of all enrichment technologies, is based on the fact that laser beams can sometimes selectively excite individual isotopes. Excited isotopes enter into chemical reactions that allow them to be separated from other isotopes of the same element. If laser enrichment technology becomes well developed and widespread, the impact could be enormous. The energy requirements are comparatively slight; little space is required; and the cost will be trivial. Laser enrichment could make weapons-grade material available to any government and to any determined organization.

Nobel Laureate Hans Bethe, a forceful advocate of commercial nuclear power, has expressed the hope that, when developed, laser enrichment technology will be kept secret for as long as possible—perhaps even twenty or thirty years. However, this seems to be wishful thinking. Ground-breaking work has already been done in several countries, and important advances in the field of laser enrichment are even now being reported in unclassified publications.

5. The most lucid discussion I have seen of the radiation issue is in J. Martin Brown, "Health, Safety, and Social Issues of Nuclear Power," in W. C. Reynolds, ed., *The California Nuclear Initiative: Analysis and Discussion of the Issues* (Institute for Energy Studies, Stanford University, 1976).

6. The standard reference in this difficult area is by the Advisory Committee on the Biological Effects of Ionizing Radiation (BEIR), *The Effects on Populations of Exposure to Low Levels of Ionizing Radiation* (Washington, D.C.: National Academy of Sciences, 1972).

7. Zhores Medvedev, "Two Decades of Dissidence," *New Scientist*, Nov. 4, 1976.

8. Testimony by Henry Eschwage, director of the Resources and Economic Development Division of the U.S. General Accounting Office, before the Subcommittee on Conservation, Energy, and Natural Resources of the House Committee on Government Operations, Feb. 23, 1976.

9. U.S. General Accounting Office, "Improvements in the Land Disposal of Radioactive Waste—A Problem of Centuries," 1976.

10. Norma Turner, "Nuclear Waste Drop in the Ocean," *New Scientist,* Oct. 30, 1975; *Weekly Environment Report,* June 23, 1975.

11. Irwin C. Bupp et al., "The Economics of Nuclear Power," *Technology Review,* Feb., 1975.

12. Amory B. Lovins, *Scale, Centralization, and Electrification in Energy Systems,* paper prepared for a Symposium on Future Strategies of Energy Development, at Oak Ridge Associated Universities, Tenn., Oct. 20–21, 1976.

13. The Nuclear Energy Agency of the Organization for Economic Cooperation and Development, and the International Atomic Energy Agency, *Uranium: Resources, Production, and Demand (1975);* similar estimates are provided by chap. 7, "Nuclear Resources," in *World Energy Conference Survey of Energy Resources* (New York: World Energy Conference, 1974); Robert D. Nininger, "Uranium Resources," ERDA statement to the House Subcommittee on Energy and Environment, June 5, 1975; and Committee on Mineral Resources and the Environment, National Academy of Sciences, *Supplementary Report: Reserves and Resources of Uranium in the United States,* ISBN 0-309-20423 (Washington, D.C.: NAS/NAC, 1976).

14. For a first-rate summary of the principal nuclear accidents to date by a thoughtful and knowledgeable nuclear critic, see Walter C. Patterson, *Nuclear Power* (Harmondsworth, England: Penguin Books, 1976).

15. When the U 235 nucleus is split, its components (92 protons and 143 neutrons) become rearranged in two smaller atoms and several subatomic particles. Less "binding energy" is needed to hold together the nuclei of the several small atoms than was needed to bind the sub-atomic particles in the one large atom. When the large atom is split, the excess binding energy is released, captured as heat in the reactor, and used to boil water.

When a U 235 atom is split into smaller atoms, neutrons not incorporated in the new elements fly off. Some of the free neutrons strike and split other U 235 atoms, causing a self-sustaining chain reaction. Many of the neutrons, however, do not encounter U 235 atoms. Some are absorbed by the moderator in the reactor core; some bombard the walls and other parts of the reactor vessel, causing them to weaken and become radioactive; and some neutrons encounter atoms of non-fissionable U 238.

Under certain circumstances, a U 238 atom will "capture" a stray neutron. This addition changes the stable U 238 atom into an unstable uranium isotope that quickly decays into plutonium 239. Plutonium 239 is itself a fissile fuel, which can be split to power a reactor, giving off free neutrons that serve to continue the chain reaction. All uranium-fueled reactors transform some U 238 into plutonium. As soon as plutonium is formed, it begins contributing to the reactor's fissions. By the time fuel is removed from a light water reactor (LWR) about half the fissions are of plutonium. A 1,000-megawatt LWR, operating at full power, will produce about 375 pounds of fissionable plutonium each year.

16. U.S. Atomic Energy Commission, "Reactor Safety Study: An Assessment of Accident Risks in U.S. Commercial Nuclear Power Plants," WASH-1400 (1975). For a critique of the techniques employed by the Rasmussen study, see Milton Kamins, *A Reliability Review of the Reactor Safety Study* (Santa Monica, Calif.: Rand Corporation, 1975).

17. Daniel F. Ford and Henry W. Kendall, *An Assessment of the Emergency Core Cooling Systems Rulemaking Hearings* (San Francisco: Union of Concerned Scientists and Friends of the Earth, 1974).

18. The cost of repairing actual fire damage was about $7 million. The cost of idle investment in the two shut-down reactors, according to the TVA public information office (April 9, 1976), was about $10 million per month.

19. Amory B. Lovins and John H. Price, *Non-Nuclear Futures: The Case for an Ethical Energy Strategy* (Cambridge, Mass.: Ballinger, 1975).

20. An exceptionally thoughtful and provocative assessment of the breeder can be found in the Royal Commission on Environmental Pollution's sixth report, *Nuclear Power and the Environment* (London: Her Majesty's Stationery Office, Sept., 1976). Another excellent, somewhat more technical reference is Thomas B. Cochran, *The Liquid Metal*

Fast Breeder Reactor: An Environmental and Economic Critique (Baltimore: Johns Hopkins Press for Resources for the Future, 1974).

21. Richard Webb, in *The Accident Hazards of Nuclear Power Plants* (Amherst: University of Massachusetts Press, 1976), calculates that the explosive force would be more than ample to destroy any plausible containment structure.

22. Leonard Ross, "How 'Atoms for Peace' Became Bombs for Sale," *New York Times Magazine*, Dec. 5, 1976. George H. Quester, "Can Proliferation Now Be Stopped?," *Foreign Affairs*, Oct., 1974; Lincoln P. Bloomfield, "Nuclear Spread and World Order," *Foreign Affairs*, July, 1975; Frank Barnaby for the Stockholm International Peace Institute, *The Nuclear Age* (Cambridge, Mass.: MIT Press, 1975); Mason Willrich, ed., *Civil Nuclear Power and International Security* (New York: Praeger, 1971). An especially provocative paper on recent international changes is "What's New on Nuclear Proliferation?," prepared by George H. Quester for the 1975 Aspen Workshop on Arms Control.

23. Committee to Study the Long-Term Worldwide Effects of Multiple Nuclear Weapons Detonation, National Academy of Sciences, *Long-Term Worldwide Effects of Multiple Nuclear Weapons Detonations* (Washington, D.C.: NAS/NAC, 1975).

24. Barry Schneider, "Big Bangs from Little Bombs," *Bulletin of the Atomic Scientists*, Sept., 1975; William Epstein, *Retrospective on the NPT Review Conference: Proposals for the Future* (Muscatine, Iowa: Stanley Foundation, 1975).

25. William Epstein, "Failure at the NPT Review Conference," *Bulletin of the Atomic Scientists*, Sept., 1975; Epstein, *Retrospective on the NPT Review Conference*.

26. A cogent critique of this "back-door" approach to nuclear weapons can be found in Alva Myrdal, " 'Peaceful' Nuclear Explosions," *Bulletin of the Atomic Scientists*, May, 1975.

27. Mason Willrich and Theodore B. Taylor, *Nuclear Theft: Risks and Safeguards* (Cambridge, Mass: Ballinger, 1974).

28. Report to Congress by the General Accounting Office, "Improvements Needed in the Program for the Protection of Special Nuclear Material" (1973).

29. An excellent international survey of the potential for nuclear terrorism is by the Mitre Corporation, "The Threat to Licensed Nuclear Facilities," MTR-7022 (1975).

30. John P. Holdren, "Hazards of the Nuclear Fuel Cycle," *Bulletin of the Atomic Scientists*, Oct., 1974.

31. This theme is carefully developed by Amory B. Lovins in "Energy Strategy: The Road Not Taken?," *Foreign Affairs*, vol. 55, no. 1 (Oct., 1976).

Chapter 4. The Case for Conservation

1. A notable exception is Sweden, where the attempt is now being made to reduce the annual growth rate of fuel use from 4.5 percent to 2 percent through 1985. By 1986, Swedish fuel use would, under this plan, reach a plateau, where it would remain indefinitely.

2. Herman E. Daly, "Energy Demand Forecasting: Prediction or Planning?," *American Institute of Planners Journal*, Jan., 1976.

3. Energy Policy Project of the Ford Foundation, *A Time to Choose* (Cambridge, Mass.: Ballinger, 1974).

4. H. E. Daly, ed., *Toward a Steady State Economy* (San Francisco: W. H. Freeman, 1973). Ezra J. Mishan, *The Costs of Economic Growth* (New York: Praeger, 1967). K. William Kapp, *Social Costs of Private Enterprise*, rev. ed. (New York: Schocken, 1971). Fred Hirsch, *Social Limits to Growth* (Cambridge, Mass.: Harvard University Press, 1976).

5. Barry Commoner, *The Poverty of Power* (New York: Knopf, 1976).

6. Bruce Hannon, "Energy, Growth, and Altruism," First Prize–winning paper at the Limits to Growth Conference, 1975.

7. Lee Schipper and A. J. Lichtenberg, *Efficient Energy Use and Well-Being: The Swedish Example* (Berkeley: Lawrence Berkeley Laboratory, April, 1976). Richard L. Goen and Ronald White, *Comparison of Energy Consumption between West*

Germany and the United States (Menlo Park: Stanford Research Institute, June, 1975).

8. John G. Myers, "Energy Conservation and Economic Growth—Are They Compatible?," *Conference Board Record*, Feb., 1975.

9. A splendid examination of the opportunities for technical conservation can be found in Lee Schipper, "Raising the Productivity of Energy Utilization," in Jack M. Hollander, ed., *Annual Review of Energy* (Palo Alto: Annual Reviews, Inc., 1976).

10. American Physical Society, *Efficient Use of Energy: A Physics Perspective* (Washington, D.C., 1975).

11. Marc H. Ross and Robert H. Williams, "Assessing the Potential for Energy Conservation" (Albany: Institute for Policy Alternatives, July 1, 1975).

12. Arjun Makhijani, *Energy Policy for the Rural Third World* (London: International Institute for Environment and Development, 1976).

13. Richard A. Walker and David B. Large, "The Economics of Energy Extravagance," *Ecology Law Quarterly*, vol. 4, no. 963, 1975 (reprint).

Chapter 5. Watts for Dinner: Food and Fuel

1. David Pimental et al., "Food Production and the Energy Crisis," *Science*, Nov. 2, 1973.

2. Eric Hirst, "Energy Use for Food," ORNL-NSF-EP-57, Oak Ridge National Laboratory, Oct., 1973.

3. Lester R. Brown, with Erik Eckholm, *By Bread Alone* (New York: Praeger, 1974).

4. Lester R. Brown, *The Politics and Responsibility of the North American Breadbasket*, Worldwatch Paper No. 2 (Washington, D.C.: Worldwatch Institute, Oct., 1975).

5. David Pimental et al., "Land Degradation: Effects on Food and Energy Resources," *Science*, vol. 194, no. 4261 (Oct. 8, 1976).

6. William Lockeretz et al., "Organic and Conventional Crop Production in the Corn Belt" (St. Louis: Center for the Biology of Natural Systems at Washington University, June, 1976).

7. Jonathan Allen, "Sewage Farming," *Environment*, vol. 15, no. 3 (April, 1973).

8. Carol and John Steinhart, *Energy: Sources, Use, and Role in Human Affairs* (North Scituate, Mass.: Duxbury Press, 1974).

9. Arjun Makhijani, in collaboration with Alan Poole, *Energy and Agriculture in the Third World* (Cambridge, Mass.: Ballinger, 1975).

10. New York *Times*, Aug. 25, 1976.

11. Gerald Leach, *Energy and Food Production* (London: International Institute for Environment and Development, 1975).

12. Roger Revelle, "Energy Use in Rural India," *Science*, June 4, 1976.

13. Arjun Makhijani, "Solar Energy and Rural Development for the Third World," *Bulletin of the Atomic Scientists*, June, 1976.

14. Keith Griffin, *Land Concentration and Rural Poverty* (New York: Holmes & Meier, 1976). Edgar Owens and Robert Shaw, *Development Reconsidered* (Lexington, Mass.: Lexington Books, 1972).

Chapter 6. Energy and Transportation

1. Richard L. Goen and Ronald K. White, *Comparison of Energy Consumption between West Germany and the United States* (Menlo Park: Stanford Research Institute, June, 1975).

2. Real Estate Research Corporation, *The Costs of Sprawl*, GPO 4111–00023 (Washington, D.C., April, 1974).

3. Wilfred Owen, *Transportation, Energy, and Community Design* (Washington, D.C.: International Institute for Environment and Development, March, 1975).

4. Emma Rothschild, *Paradise Lost: The Decline of the Auto-Industrial Age* (New York: Vintage Books, 1974).

5. American Physical Society, *Efficient Use of Energy: A Physics Perspective* (Washington, D.C., 1975).

6. Eric Hirst, *Energy Use for Bicycling*, Oak Ridge National Laboratory, Feb., 1974. Nina Dougherty and William Lawrence, *Bicycle Transportation*, U.S. Environmental Protection Agency, Dec., 1974.

7. Lee Pratsch, *Carpool and Buspool Matching Guide*, 4th ed., U.S. Department of Transportation, Jan., 1975.

8. Bradford C. Snell, *American Ground Transport*, presented to the U.S. Senate Committee on the Judiciary, Feb., 1974.

9. Richard A. Rice, "System Energy and Future Transportation," *Technology Review*, Jan., 1972.

10. Eric Hirst, *Energy Intensiveness of Passenger and Freight Transport Modes: 1950–1970*, Oak Ridge National Laboratory, April, 1973.

11. Tom Alexander, "A New Outbreak of Zeppelin Fever," *Fortune*, Dec., 1973.

12. J. King, "Appropriate Technology in Shipping: Wind Powered Ships," Department of Naval Architecture and Shipbuilding, University of Newcastle upon Tyne.

13. Eric Hirst, "Transportation Energy Conservation Policies," *Science*, April 2, 1976.

Chapter 7. Btu's and Buildings: Energy and Shelter

1. American Institute of Architects, *Energy and the Built Environment* (Washington, D.C., 1975). American Institute of Architects, *A Nation of Energy Efficient Buildings by 1990* (Washington, D.C., 1975).

2. Richard Stein, "A Matter of Design," *Environment*, Oct., 1972.

3. Owens-Corning Fiberglass, *Energy-Saving Homes: The Arkansas Story*, June, 1976.

4. The Real Estate Research Corporation, *The Costs of Sprawl*, GPO 4111–00023 (Washington, D.C., April, 1974).

5. John A. Duffie and William A. Beckman, "Solar Heating and Cooling," *Science*, Jan. 16, 1976. A fine photographic survey of several U.S. solar homes can be found in Norma Skurka and Jon Naar, *Design for a Limited Planet* (New York: Ballantine, 1976). A more comprehensive survey is W. A. Shurcliff, *Solar Heated Buildings: A Brief Survey*, 13th ed. (San Diego: Solar Energy Digest, 1977). Active approaches to solar heating are described in W. A. Shurcliff, "Active-Type Solar Heating Systems for Houses: A Technology in Ferment," *Bulletin of the Atomic Scientists*, Feb., 1976. Passive solar design is explained in Raymond W. Bliss, "Why Not Just Build the House Right in the First Place?" *Bulletin of the Atomic Scientists*, March, 1976, and by Bruce Anderson, "Low Impact Solutions," *Solar Age*, Sept., 1976.

6. Steve Baer, *Sunspots* (Albuquerque, N.M.: Zomeworks, 1975).

7. M. Telkes, "Thermal Storage in Sodium Thiosulfate Pentahydrate," presented to Intersociety Energy Commission Engineering Conference, University of Delaware, Aug. 18, 1975.

8. H. C. Fischer, ed., *Summary of the Annual Cycle Energy System Workshop I* (Oak Ridge, Tenn.: Oak Ridge National Laboratory, July, 1976).

9. Complete information on and specifications for this air-conditioning system are available from the Yazaki Buhin Company, Ltd., 390, Umeda Kosai City, Shizuoka Prefecture, Japan.

10. The Mitre Corporation, *An Economic Analysis of Solar Water and Space Heating* (Washington, D.C.: Energy Research and Development Administration, Nov., 1976); William D. Schulze et al., *The Economics of Solar Home Heating*, prepared for the Joint Economic Committee of the U.S. Congress (Washington, D.C.: U.S. Government Printing Office, March 13, 1977).

11. This is the most persuasive argument available to those who favor utility investments in solar technologies and conservation. A utility should in theory be willing to make electricity-saving investments up to the high *marginal* cost of new power plants, whereas the consumer will want to make only those investments that are sensible in light of *average*

electrical bills. Arguments against such utility involvement are generally based on the assumption that the utility will charge high prices for equipment and labor while demanding an exorbitant rate of return on its investment. In parts of the world where utilities are government regulated, this argument loses much of its force.

12. Frost & Sullivan, *The U.S. Solar Power Market,* Report No. 348, New York, 1975. Frost & Sullivan estimates the total annual U.S. solar market for 1985, including wind power and biomass, at $10 billion. In its *A Nation of Energy-Efficient Buildings by 1990* (Washington, D.C., 1975), the American Institute of Architects calculates that an ambitious program of conservation and solar development could save the United States the equivalent of 12.5 million barrels of oil a day in 1990. The institutional obstacles such rapid solar development would have to overcome are discussed in R. Schoen, A. S. Hirshberg, and J. Weingart, *New Energy Technology for Buildings* (Cambridge, Mass.: Ballinger, 1975).

Chapter 8. Energy and Economic Growth

1. J. K. Galbraith, *The Affluent Society* (Boston: Houghton Mifflin, 1958).
2. John G. Myers et al., *Energy Consumption in Manufacturing: Report to the Energy Policy Project* (Cambridge, Mass.: Ballinger, 1974).
3. Nicholas Georgescu-Roegen, *The Entropy Law and the Economic Process* (Cambridge, Mass.: Harvard University Press, 1971).
4. U.S. Federal Energy Administration, *Comparison of Energy Consumption between West Germany and the United States,* Conservation Paper No. 33, Washington, D.C., 1976.
5. Charles A. Berg, "Conservation in Industry," *Science,* April 19, 1974.
6. Charles A. Berg, "Potential for Energy Conservation in Industry," in Jack M. Hollander, ed., *Annual Review of Energy* (Palo Alto: Annual Reviews, Inc., 1976).
7. The next several examples are drawn from E. P. Gyftopoulos, L. J. Lazaridis, and T. F. Widmer, *Potential Fuel Effectiveness in Industry,* report to the Energy Policy Project of the Ford Foundation (Cambridge, Mass.: Ballinger, 1974).
8. Dow Chemical Company, Environmental Research Institute of Michigan, Townsend-Greenspan & Co., and Cravith, Swaine & Moore, *Energy Industrial Center Study,* report to the National Science Foundation (OEP74–20242), 1975.
9. Richard Grossman and Gail Daneker, *Jobs and Energy* (Washington, D.C.: Environmentalists for Full Employment, 1977).
10. Richard Stein, "A Matter of Design," *Environment,* Oct., 1972.
11. S. L. Blum, "Tapping Resources in Municipal Solid Waste," *Science,* vol. 191, no. 4228 (Feb. 20, 1976).
12. R. Stephen Berry, "Reducing the Energy Demand," *New York Times,* Feb. 12, 1976.
13. Boyce Rensberger, "Coining Trash," *New York Times Magazine,* Dec. 7, 1975.
14. Barry Stein, Testimony before the Senate Select Committee on Small Business, Dec. 2, 1975.

Chapter 9. Turning toward the Sun

1. By far the largest fraction of current commercial solar usage is of human biomass. In many Third World countries, firewood, dung, and crop residues constitute 90 percent of all energy use. Calculations regarding the magnitude of this usage can be found in Arjun Makhijani and Alan Poole, *Energy and Agriculture in the Third World* (Cambridge, Mass.: Ballinger, 1975), and D. F. Earl, *Forest Energy and Economic Development* (Oxford: Clarendon Press, 1975). Hydropower ranks next, providing more than one-fifth of all electricity and about 3 percent of all end-use energy. See United Nations, *World Energy Supplies: 1950–1974* (New York: Department of Economic and Social Affairs, 1976).
2. Insight into the many vital but unnoticed functions performed for humankind by

the sun can be gleaned from Frank von Hippel and Robert H. Williams, "Solar Technologies," *Bulletin of the Atomic Scientists,* Nov., 1975, and Steve Baer, "Clothesline Paradox," *Elements,* Nov., 1975. The temperature estimate for a sunless earth was provided in Vincent E. McKelvey, "Solar Energy in Earth Processes," *Technology Review,* April, 1975.

3. John V. Krutilla and R. Talbot Page, "Energy Policy from an Environmental Perspective," in Robert J. Kalter and William A. Vogely, eds., *Energy Supply and Government Policy* (Ithaca, N.Y.: Cornell University Press, 1976); John S. Reuyl et al., *A Preliminary Social and Environmental Assessment of the ERDA Solar Energy Program, 1975–2020,* Vols. I and II (Menlo Park, Calif.: Stanford Research Institute, 1976), found solar technologies to be environmentally attractive compared to the alternatives.

4. Hans H. Landsberg, "Low-Cost Abundant Energy: Paradise Lost?" (Washington, D.C.: Resources for the Future Reprint No. 112, Dec., 1973).

5. The U.S. Federal Energy Administration publishes a semiannual *Survey of Solar Collector Manufacturing Activity;* the 1977 estimate is by Ronald Peterson, director of Grummon Energy Systems, one of the largest manufacturers of solar collectors.

6. Largely because conventional fuels pose costly transportation and distribution problems in remote areas, the largest immediate market for expensive photovoltaic cells may, strangely enough, be in the world's poorest countries. Charles Weiss and Simon Pak, "Developing Country Applications of Photovoltaic Cells," presented to the ERDA National Solar Photovoltaic Program Review Meeting, San Diego, Calif., Jan. 20, 1976.

7. M. L. Baughman and D. J. Bottaro, *Electric Power Transmission and Distribution Systems: Costs and Their Allocation* (Austin: University of Texas Center for Energy Studies, July, 1975).

8. An excellent exploration of the concept of thermodynamic matching is in "Efficient Use of Energy: A Physics Perspective," American Physical Society, Jan., 1975 (reprinted in U.S. House of Representatives, Committee on Science and Technology, Part I, ERDA Authorization Hearings, Feb. 18, 1975). Simpler explanations can be found in Barry Commoner, *The Poverty of Power* (New York: Knopf, 1976), and Denis Hayes, *Energy: The Case for Conservation* (Washington, D.C.: Worldwatch Institute, Jan., 1976).

9. Amory B. Lovins, "Scale, Centralization, and Electrification in Energy Systems," presented to a Symposium on Future Strategies of Energy Development, Oak Ridge, Tenn., Oct. 20–21, 1976. The Canadian data is in "Exploring Energy-Efficient Futures for Canada," *Conserver Society Notes,* May-June, 1976.

10. These issues are thoughtfully explored in John S. Reuyl et al., *A Preliminary Social and Environmental Assessment of the ERDA Solar Energy Program, 1975–2020;* Amory B. Lovins, "Energy Strategy: The Road Not Taken?," *Foreign Affairs,* Oct., 1976; and less directly by Rufus E. Miles, Jr., *Awakening from the American Dream: The Social and Political Limits to Growth* (New York: Universe Books, 1976); Bruce Hannon, "Energy, Land, and Equity," presented to the 41st North American Wildlife Conference, Washington, D.C., March 21–25, 1976; and E. F. Schumacher, *Small Is Beautiful: Economics as if People Mattered* (New York: Harper and Row, 1973); and William Ophuls, *Ecology and the Politics of Scarcity* (San Francisco: W. H. Freeman, 1977).

11. Among their other virtues, flat-plate collectors have a high net energy yield. A conventional collector will deliver enough energy in less than one year to pay back the energy used in its manufacture. Moreover, if collectors are recycled, the energy requirements are reduced dramatically. See the various statements on net energy in *Solar News and Views* (International Solar Energy Society, American Section, Richmond, Calif.) Jan. and April, 1976.

12. W. A. Shurcliff, *Solar Heated Buildings: A Brief Survey,* 13th ed. (San Diego: Solar Energy Digest, 1977). See also Philip Steadman, *Energy, Environment, and Building* (Philadelphia: Academy of Natural Sciences of Pennsylvania, 1975).

13. Multiple-effect solar stills are described in "Solar Desalting Process Breakthroughs," *Solar Energy Digest,* June, 1976.

14. "French Solar-Powered Irrigation Pump Installed in Mexico," *Solar Energy Digest,* Feb., 1976.

15. D. Proctor and R. F. White, "The Application of Solar Energy in the Food Processing Industry," presented to a meeting of the Australian and New Zealand Sections of ISES, Melbourne, Australia, July 2, 1975.

16. Malcolm Fraser, *Analysis of the Economic Potential of Solar Thermal Energy to Provide Industrial Process Heat* (Warrenton, Va.: Intertechnology Corp., 1977). A concentrating solar collector can quite easily obtain a temperature of 288°C.

17. The energy demand projections used by the U.S. Energy Research and Development Administration to justify a massive nuclear power program were carefully analyzed by Frank von Hippel and Robert Williams, "Energy Waste and Nuclear Power Growth," *Bulletin of the Atomic Scientists,* Dec., 1976. The authors found that the projections demanded the use of electricity for virtually everything. The most egregious example of electrical "padding" was for industrial process heat. Virtually no electricity is used this way today; yet the projections show the 2020 electrical demand for process heat to be larger than all electricity used throughout the entire U.S. economy in 1975. Fraser, in *Analysis of the Economic Potential,* found that half of this energy could be provided by direct solar heating; most of the remaining half can be more easily met with biomass or other fuels than with electricity.

18. The utility, the Public Service Company of New Mexico, hopes to make the first such hybrid conversion of its Person power plant in Albuquerque.

19. Aden Baker Meinel and Marjorie Pettit Meinel, *Power for the People* (Tucson, Ariz.: privately published, 1970).

20. Arguments for closed-cycle OTECs can be found in U.S. House of Representatives, Subcommittee on Energy of the Committee on Science and Astronautics, *Solar Thermal Power,* Hearings, May 23, 1974. Open-cycle OTECs are advocated in Earl J. Beck, "Ocean Thermal Gradient Hydraulic Power Plant," *Science,* July 25, 1975, and Clarence Zener and John Fetkovitch, "Foam Solar Sea Power Plant," *Science,* July 25, 1975.

21. An excellent series of papers was prepared under the auspices of the American Society of International Law for the 1976 Workshop on Legal, Political, and Institutional Aspects of Ocean Thermal Energy Conversion. For a more optimistic assessment of OTEC economics, see Clarence Zener, "Solar Sea Power," *Bulletin of the Atomic Scientists,* Jan., 1976, or George Haber, "Solar Power from the Oceans," *New Scientist,* March 10, 1977.

22. R. H. Williams, "The Greenhouse Effect for Ocean Based Solar Energy Systems," Working Paper No. 21, Center for Environmental Studies, Princeton University, Oct., 1975.

23. An excellent introduction to photovoltaics can be found in Bruce Chalmers, "The Photovoltaic Generation of Electricity," *Scientific American,* Oct., 1976. For a more detailed treatment, see Joseph A. Merrigan, *Sunlight to Electricity: Prospects for Solar Energy Conversion by Photovoltaics* (Cambridge, Mass.: MIT Press, 1975).

24. A recent technical survey of photovoltaic materials and techniques can be found in the two-volume *Proceedings of the E.R.D.A. Solar Photovoltaic Program Review Meeting,* Aug. 3–6, 1976 (Springfield, Va.: National Technical Information Service, 1976).

25. See, for example, the testimony of Paul Rappaport and others in *Solar Photovoltaic Energy.* Hearings before the Subcommittee on Energy of the House Committee on Science and Astronautics, Washington, D.C., June 6 and 11, 1974.

26. A useful overview of the Japanese program is provided by Akira Uehara, "Solar Energy Research and Development in Quest for New Energy Sources," *Technocrat,* vol. 9, no. 3. See also *Japan's Sunshine Project* (Tokyo: MITI Agency of Industrial Science and Technology, 1975).

27. The two-year payback period (for cells with an expected lifetime of more than twenty years) has become conventional wisdom among the silicon photovoltaic specialists. See, for example, Martin Wolfe, "Methods for Low-Cost Manufacture of Integrated Solar Arrays," and P. A. Iles, "Energy Economics in Solar Cell Processing," both in *Proceedings of the Symposium on the Material Science Aspects of Thin Film Systems for Solar Energy Conversion,* May 20–24, Tucson, Ariz. (Washington, D.C.: National Science Foundation,

1974). The calculations by Slesser and Hounam based upon a two-year payback are in M. Slesser and I. Hounam, "Solar Energy Breeders," *Nature*, July 22, 1976. E. L. Ralph, vice-president for research at Spectrolab, claims that his company's cells now have a payback period of 87 days, and that the theoretical minimum would be on the order of 30 hours, according to a personal communication with Dr. Peter Glaser of Arthur D. Little.

28. Photovoltaics could, of course, also be used in highly centralized arrays in areas of high insolation. The advantages of decentralization are more social than technical. At the extreme are proposals to obtain large amounts of energy from photovoltaic cells on orbiting satellites, with the energy beamed down to earth via microwaves. The idea was first suggested by Peter Glaser, "Power from the Sun: Its Future," *Science*, Nov., 1968, and has more recently been popularized by Gerald K. O'Neil, "Space Colonization and Energy Supply to the Earth," *Co-Evolution Quarterly*, Fall, 1975. The concept appears to have no insurmountable technical flaws, but is of dubious desirability. Simple, decentralized terrestrial uses of photovoltaics have far more to recommend them.

29. Some argue that storing energy from renewable sources would require people to change their life styles to conform to the periodicity of such sources. However, life-style adjustments would attend a switch to nuclear as well as solar substitutes for oil and gas.

Storage costs will motivate users of solar energy sources to schedule their energy-using activities for daylight hours. Similarly, the cost of storing nuclear power will encourage consumers to even out their daily energy use. In neither case is the requirement absolute, but in both people will be rewarded for tailoring their demands to fit supplies.

The Sørensen estimate can be found in Bent Sørensen, "Dependability of Wind Energy Generators with Short-Term Energy Storage," *Science*, Nov. 26, 1976.

30. Public Service Electric and Gas Company of New Jersey, *An Assessment of Energy Storage Systems for Use by Electric Utilities* (Palo Alto, Calif.: Electric Power Research Institute, 1976).

31. Clark, *Energy for Survival.*

32. A comprehensive recent article by some of the foremost proponents of a "hydrogen economy" is D. P. Gregory and J. B. Parghorn, "Hydrogen Energy," in Jack M. Hollander, ed., *Annual Review of Energy*; a somewhat more skeptical appraisal is in J. K. Dawson, "Prospects for Hydrogen as an Energy Source," *Nature*, June 21, 1974. An excellent quarterly technical journal, the *International Journal of Hydrogen Energy*, is available through Pergamon Press, Ltd., Oxford, U.K.

The fuel cell is a device that produces electricity directly from fuel through electrochemical reactions. Invented in 1839 by Sir William Grove, the fuel cell has been put to practical use in the space program. The United Technologies Corporation has embarked upon a $42 million research effort to develop a commercial fuel cell, and U.S. utilities have already signed options for the first 56 units produced. Vigorous research is also under way in many other countries.

Fuel cells have several major advantages over conventional technologies. They involve no combustion and hence virtually no pollutants. Sixty percent conversion efficiencies are common, and 75 percent efficiencies have been reported. Unlike conventional power plants, fuel cells are as efficient with a partial load as with a peak load. Moreover, their modular design shortens construction lead times, since new modules can be added as needed. The use of decentralized fuel cells would also save expenses of constructing long-distance power lines from huge generating facilities, and would allow waste heat to be productively employed. Fuel cells are quiet, and they conserve water.

33. H. C. Herbst, "Air Storage–Gas Turbine: A New Possibility of Peak Current Production," *Proceedings of the Technical Conference on Storage Systems for Secondary Energy*, Stuttgart, Oct., 1974. For a broad overview of this technology, see also A. J. Giramonti and R. D. Lessard, "Exploratory Evaluation of Compressed Air Storage Peak-Power Systems," *Energy Sources*, vol. 1, no. 3, 1974; and D. L. Ayers and D. R. Hoover, "Gas Turbine Systems Using Underground Compressed Air Storage," presented at the American Power Conference, Chicago, April 29–May 1, 1974.

34. Fritz R. Kalhammer and Thomas R. Schneider, "Energy Storage," in Jack M. Hollander, ed., *Annual Review of Energy*. See also Julian McCaull, "Storing the Sun," *Environment*, June, 1976. Much interesting material concerning flywheels and other storage devices can be found in J. M. Savino, ed., *Wind Energy Conversion Systems*, First Workshop Proceedings (Washington, D.C.: National Science Foundation, Dec., 1973).

35. A good survey of current battery prospects is in Kalhammer and Schneider, "Energy Storage." An interesting new battery idea is described in M. S. Whittingham, "Electrical Energy Storage and Intercalation Chemistry," *Science*, June 11, 1976.

36. Comprehensive overviews of solar energy can be found in Farrington Daniels, *Direct Use of the Sun's Energy* (New York: Ballantine, 1974), and B. J. Brinkworth, *Solar Energy for Man* (New York: Wiley, 1972). Two more recent articles in *Technology Review* provide excellent analyses of the solar potential: Walter E. Morrow, Jr., "Solar Energy: Its Time Is Near," Dec., 1973, and John B. Goodenough, "The Options for Using the Sun," Oct.-Nov., 1976. The most exhaustive survey of all renewable energy technologies remains Wilson Clark, *Energy for Survival* (Garden City, N.Y.: Anchor Press/Doubleday, 1974). A recent survey of U.S. corporate interest in several of these technologies is Stewart W. Herman and James S. Cannon, *Energy Futures* (New York: Inform, Inc., 1976). An overview of current international solar research efforts can be found in F. de Winter and J. W. de Winter, eds., *Description of the Solar Energy R & D Programs in Many Nations* (Santa Clara, Calif.: Atlas Corp., Feb., 1976). And an extremely readable introduction to many of the world's leading solar researchers, and to the technologies they are developing, is Daniel Behrman, *Solar Energy: The Awakening Science* (Boston: Little, Brown, 1976).

Chapter 10. Wind and Water Power

1. The largest of these sailing vessels captured about four megawatts of power from the wind. I am indebted to Professor Frank von Hippel of Princeton University for several ideas in this chapter.

2. Surveys of the history of wind power can be found in Volta Torrey, *Wind Catchers* (Brattleboro, Vt.: Stephen Green Press, 1976); E. W. Golding, *The Generation of Electricity by Wind Power* (New York: Philosophical Library, 1955); John Reynolds, *Windmills and Watermills* (New York: Praeger, 1970); A. T. H. Gross, *Wind Power Usage in Europe* (Springfield, Va.: National Technical Information Service, 1974).

3. Palmer C. Putnam, *Power from the Wind* (New York: Van Nostrand, 1948).

4. Don Hinrichsen and Patrick Cawood, "Fresh Breeze for Denmark's Windmills," *New Scientist,* June 10, 1976.

5. Stewart W. Herman and James S, Cannon, *Energy Futures* (New York: Inform, Inc., 1976); see also Marshal F. Merriam, "Wind Energy for Human Needs," *Technology Review*, Jan., 1977.

6. Frank Eldridge, *Wind Machines* (Washington, D.C.: U.S. Government Printing Office, 1976); J. M. Savino, ed., *Wind Energy Conversion Systems: First Workshop Proceedings* (Washington, D.C.: U.S. Government Printing Office, 1973); *Wind Energy:* Hearing before the Subcommittee on Energy of the U.S. House Committee on Science and Astronautics (Washington, D.C.: U.S. Government Printing Office, May 21, 1974).

7. J. A. Potworowski and B. Henry, "Harnessing the Wind," *Conserver Society Notes,* Fall, 1976. The cost estimate is from R. S. Rangi et al., "Wind Power and the Vertical-Axis Wind Turbine Developed at the National Research Council," DME/NAE *Quarterly Bulletin*, No. 1974(2). A good introduction to the Darrieus turbine can be found in B. B. Blackwell and L. F. Feltz, "Wind Energy: A Revitalized Pursuit" (Albuquerque, N.M.: Sandia Laboratories, March, 1975).

8. J. T. Yen, "Tornado-Type Wind Energy Systems: Basic Considerations," presented to the International Symposium on Wind Energy Systems, St. John's College, Cambridge, England, Sept. 7–9, 1976.

9. Cost estimates can be found in Federal Energy Administration, *Project Independence Final Task Force Report on Solar Energy* (Washington, D.C.: U.S. Government

Printing Office, 1974); somewhat more optimistic estimates are in David R. Inglis, "Wind Power Now!," *Bulletin of the Atomic Scientists*, Oct., 1975, and Bent Sørensen, "Wind Energy," *Bulletin of the Atomic Scientists*, Sept., 1976.

10. Edward N. Lorentz, *The Nature and Theory of the General Circulation of the Atmosphere* (Geneva: World Meteorological Organization, 1967).

11. A small fraction of the planet's wind produces some 150 million square miles of ocean waves. Britain's Department of Energy is spending a million dollars a year on experimental efforts to tap the waves that constantly break along Britain's long, stormy coasts. Smaller fledgling programs are under way elsewhere too, notably in Japan and the United States. More than a hundred different mechanical and hydraulic wave power devices have been proposed. Mechanical devices include the lopsided "ducks" designed by Stephen Salter of Edinburgh to obtain the maximum possible rock from passing waves, and the strings of contouring rafts, which work on the same principle, that Christopher Cockerell (the inventor of the Hovercraft) has proposed. The Japanese use an inverted box to capture wave energy hydraulically. When waves rise, air is pushed out of holes in the top of the box; as the wave falls, air is sucked in. These air currents are now used to power Japanese navigation buoys, and strings of such boxes may well be multiplied into power sources of commercial value in the near future. See S. H. Salter, "Wave Power," *Nature*, June 21, 1974, and Michael Kenward, "Waves a Million," *New Scientist*, May 6, 1976.

12. Hydropower resource estimates are clouded by uncertain data and ambiguous definitions. For example, such estimates typically measure either the maximum generating capacity that is usable 95 percent of the year or else the capacity usable under conditions of average annual flow. Although these figures can differ by as much as 300 percent, those who make hydropower assessments often fail to state which figure they are using.

This paper employs the more conservative 95 percent figure and then reduces it sharply to reflect new constraints being imposed by environmental and agricultural interests, and also to reflect the futility of damming silt-laden streams that drain geologically unstable areas.

The most comprehensive of the conventional hydropower resource estimates can be found in World Energy Conference, *Survey of Energy Resources* (New York: privately published for the World Energy Conference, 1974).

13. Fine surveys of small-scale hydropower technologies appear in Robin Saunders, "Harnessing the Power of Water," *Energy Primer* (Menlo Park: Portola Institute, 1974), and Ken Darrow and Rick Pam, *Appropriate Technology Sourcebook* (Stanford, Calif.: Volunteers in Asia Press, 1976). An intriguing approach to "low head" hydroelectricity is discussed in Yvonne Howell, "New Straight-Flow Turbine," *Sunworld* (published quarterly by the International Solar Energy Society), Feb., 1977.

14. Peter H. Freeman, *Large Dams and the Environment: Recommendations for Development Planning* (Washington, D.C.: International Institute for Environment and Development, March, 1977).

15. Vaclav Smil, "Intermediate Technology in China," *Bulletin of the Atomic Scientists*, Feb., 1977.

16. Erik Eckholm, *Losing Ground: Environmental Stress and World Food Prospects* (New York: Norton, 1976). See also *Ambio*, Special Issue on Water, vol. 6, no. 1, 1977.

Chapter 11. Plant Power: Biological Sources of Energy

1. H. Lieth and R. H. Whittaker, eds., *Primary Productivity of the Biosphere* (New York: Springer-Verlag, 1975); E. E. Reichle, J. F. Franklin, and D. W. Goodall, eds., *Productivity of World Ecosystems* (Washington, D.C.: National Academy of Sciences, 1975); E. E. Robertson and H. M. Lapp, "Gaseous Fuels," in *Proceedings of a Conference on Capturing the Sun through Bioconversion* (Washington, D.C.: Washington Center for Metropolitan Studies, 1976).

2. Alan Poole and Robert H. Williams, "Flower Power: Prospects for Photosynthetic

Energy," *Bulletin of the Atomic Scientists*, May, 1976; Arjun Makhijani and Alan Poole, *Energy and Agriculture in the Third World* (Cambridge, Mass.: Ballinger, 1975).

3. P. E. Henderson, *India: The Energy Sector* (Washington, D.C.: World Bank, 1975).

4. Roger Revelle, "Energy Use in Rural India," *Science*, June 4, 1976.

5. W. J. Jewell, H. R. Davis, et al., *Bioconversion of Agricultural Wastes for Pollution Control and Energy Conservation* (Ithaca, N.Y.: Cornell University Press, 1976).

6. Poole and Williams, "Flower Power."

7. R. H. Whittaker and G. M. Woodwell, *Productivity of Forest Ecosystems* (Paris: UNESCO, 1971).

8. D. F. Earl, *Forest Energy and Economic Development.*

9. Erik Eckholm, *The Other Energy Crisis: Firewood* (Washington, D.C.: World-watch Institute, 1975).

10. J. S. Bethel and G. F. Schreuder, "Forests Resources: An Overview," *Science*, Feb. 20, 1976.

11. S. B. Richardson, *Forestry in Communist China* (Baltimore: Johns Hopkins Press, 1966).

12. G. C. Szego and C. C. Kemp, "The Energy Plantation," U.S. House of Representatives, Subcommittee on Energy of the Committee on Science and Astronautics, Hearings, June 13, 1974.

13. J. B. Grantham and T. H. Ellis, "Potentials of Wood for Producing Energy," *Journal of Forestry*, vol. 72, no. 9, 1974.

14. Melvin Calvin, "Hydrocarbons via Photosynthesis," presented to the 110th Rubber Division Meeting of the American Chemical Society, San Francisco: Oct. 5–8, 1976. Available from the American Chemical Society.

15. J. A. Alich and R. E. Inman, *Effective Utilization of Solar Energy to Produce Clean Fuel* (Menlo Park: Stanford Research Institute, 1974).

16. B. C. Wolverton, R. M. Barlow, and R. C. McDonald, *Application of Vascular Aquatic Plants for Pollution Removal, Energy and Food Production in a Biological System* (Bay St. Louis, Miss.: NASA, 1975); B. C. Wolverton, R. C. McDonald, and J. Gordon, *Bioconversion of Water Hyacinths into Methane Gas: Part I* (Bay St. Louis, Miss.: NASA, 1975). The report voicing skepticism about the U.S. potential is A. C. Robinson, J. H. Gorman, et al., *An Analysis of Market Potential of Water Hyacinth–Based Systems for Municipal Wastewater Treatment* (Columbus, Ohio: Batelle Laboratories, 1976).

17. H. A. Wilcox, "Ocean Farming," in *Capturing the Sun through Bioconversion.* For a less sanguine appraisal of the large ocean-farm concept, see John Ryther's remarks in the same volume.

18. J. T. Pfeffer, "Reclamation of Energy from Organic Refuse: Anaerobic Digestion Processes," presented to the Third National Congress on Waste Management and Resource Recovery, San Francisco, 1974; Alan Poole, "The Potential for Energy Recovery from Organic Wastes," in R. H. Williams, ed., *The Energy Conservation Papers* (Cambridge, Mass.: Ballinger, 1975). A good annotated bibliography of do-it-yourself books on biogas plants appears in Ken Darrow and Rick Pam, *Appropriate Technology Sourcebook* (Stanford, Calif.: Volunteers in Asia Press, 1976).

19. Vaclav Smil, "Intermediate Technology in China," *Bulletin of the Atomic Scientists*, Feb., 1977.

20. *Report to the Preparatory Mission on Bio-gas Technology and Utilization.*

21. Ram Bux Singh, *Bio-Gas Plant* (Ajitmal, Etawah, India: Gobar Research Station Publication, 1971).

22. Poole and Williams, "Flower Power."

23. C. R. Prasad, K. K. Prasad, and A. K. N. Reddy, "Biogas Plants: Prospects, Problems, and Tasks," *Economic and Political Weekly*, vol. 9, no. 32–34, 1974.

24. R. N. Morse and J. R. Siemon, *Solar Energy for Australia: The Role of Biological Conversion*, presented to the Institution of Engineers, Australia, 1975.

25. G. C. Floueke and P. H. McGauhey, "Waste Materials," in Jack M. Hollander, ed., *Annual Review of Energy*, vol. 1 (Palo Alto, Calif.: Annual Reviews, Inc., 1976).

Index